Lecture Notes in Statistics 132

Edited by P. Bickel, P. Diggle, S. Fienberg, K. Krickeberg,
I. Olkin, N. Wermuth, S. Zeger

Springer
*New York
Berlin
Heidelberg
Barcelona
Budapest
Hong Kong
London
Milan
Paris
Singapore
Tokyo*

Douglas Nychka
Walter W. Piegorsch
Lawrence H. Cox (Editors)

Case Studies in Environmental Statistics

 Springer

Douglas Nychka
Department of Statistics
North Carolina State University
Raleigh, NC 27695-8203

Walter W. Piegorsch
Department of Statistics
University of South Carolina
Columbia, SC 29208

Lawrence H. Cox
U.S. Environmental Protection Agency
National Exposure Research Laboratory (MD-75)
Research Triangle Park, NC 27711

CIP data available.
Printed on acid-free paper.

© 1998 Springer-Verlag New York, Inc.
All rights reserved. This work may not be translated or copied in whole or in part without the written permission of the publisher (Springer-Verlag New York, Inc., 175 Fifth Avenue, New York, NY 10010, USA), except for brief excerpts in connection with reviews or scholarly analysis. Use in connection with any form of information storage and retrieval, electronic adaptation, computer software, or by similar or dissimilar methodology now known or hereafter developed is forbidden.
The use of general descriptive names, trade names, trademarks, etc., in this publication, even if the former are not especially identified, is not to be taken as a sign that such names, as understood by the Trade Marks and Merchandise Marks Act, may accordingly be used freely by anyone.

Camera ready copy provided by the editors.
Printed and bound by Braun-Brumfield, Ann Arbor, MI.
Printed in the United States of America.

9 8 7 6 5 4 3 2 1

ISBN 0-387-98478-X Springer-Verlag New York Berlin Heidelberg   SPIN 10668258

Preface

This book offers a set of case studies exemplifying the broad range of statistical science used in environmental studies and application. The case studies can be used for graduate courses in environmental statistics, as a resource for courses in statistics using genuine examples to illustrate statistical methodology and theory, and for courses in environmental science. Not only are these studies valuable for teaching about an essential cross-disciplinary activity but they can also be used to spur new research along directions exposed in these examples.

The studies reported here resulted from a program of research carried on by the National Institute of Statistical Sciences (NISS) during the years 1992–1996. NISS was created in 1991 as an initiative of the national statistics organizations, with the mission to renew and focus efforts of statistical science on important cross-disciplinary problems. One of NISS' first projects was a cooperative research effort with the U.S. Environmental Protection Agency (EPA) on problems of great interest to environmental science and regulation, surely one of today's most important cross-disciplinary activities.

With the support and encouragement of Gary Foley, Director of the (then) U.S. EPA Atmospheric Research and Exposure Assessment Laboratory, a project and a research team were assembled by NISS that pursued a program which produced a set of results and products from which this book was drawn. The development and success of this program of research reflect strong efforts of Peter Bloomfield, M. Ross Leadbetter, Douglas Nychka, Richard Smith, Raymond Carroll, and Douglas Simpson, as well as cooperating scientists and managers at the EPA, especially Lawrence Cox, John Creason, David Holland, Chon Shoaf, and Daniel Guth.

The many participants in the research are listed in Appendix D and denoted throughout the book. Of particular note is the effect and influence of the many NISS postdoctoral fellows who played such a highly significant role in the research, including the formulation and development of the projects. Our thanks to Feng Gao, James Hilden-Minton, Nancy McMillan, Nancy Saltzman, Laura Steinberg, Patricia Styer, and Minge Xie for their efforts.

Thanks to the editors and contributors for shaping and presenting these case studies to make them accessible for many users and for showing the way to future volumes in a series devoted to case studies and summaries of cross-disciplinary statistics research at NISS. Our thanks to John Kimmel and Springer-Verlag for creating this opportunity.

Jerome Sacks
Director, National Institute of Statistical Sciences

For more information about the authors, data sets, software and related links, refer to this volume's web companion. This electronic supplement can be accessed from the Springer homepage (www.springer-ny.com) or the National Institute of Statistical Sciences homepage (www.niss.org)

Contents

Preface		v
1	**Introduction: Problems in Environmental Monitoring and Assessment**	**1**
	Lawrence H. Cox, Douglas Nychka, Walter W. Piegorsch	
	1 Statistical Methods for Environmental Monitoring and Assessment	1
	2 Outline of Case Studies	2
	3 Sources of Data and Software	3
	Acknowledgments	4
2	**Modeling Ozone in the Chicago Urban Area**	**5**
	Jerry M. Davis, Brian K. Eder, Peter Bloomfield	
	1 Introduction	5
	1.1 Health and Environmental Effects	6
	1.2 Background	6
	1.3 Overview of the Case Studies	7
	2 Data Sources	8
	2.1 Ozone Data	8
	2.2 Meteorological Data	8
	3 Trend Analysis and Adjustment	10
	3.1 Parametric Modeling	10
	3.2 Urban Ozone in Chicago	11
	3.3 Comparison with Rural Locations	15
	4 Trends from Semiparametric Models	16
	4.1 Semiparametric Modeling	16
	4.2 Application to Chicago Urban Ozone	17
	5 Trends in Exceedances	19
	5.1 Exceedance Modeling	19
	5.2 Modeling Exceedance Probabilities for the Chicago Urban Area	20
	5.3 Modeling Excess Ozone over a Threshold	21
	5.4 Prediction of Extreme Ozone Levels	23
	6 Summary	23
	Acknowledgments	24
	References	24

3 Regional and Temporal Models for Ozone Along the Gulf Coast   27
Jerry M. Davis, Brian K. Eder, Peter Bloomfield

- 1 Introduction 27
  - 1.1 Scientific Issues 27
  - 1.2 Data and Dimension Reduction 28
- 2 Diurnal Variation in Ozone 29
  - 2.1 Singular Value Decomposition 30
  - 2.2 Urban Ozone in Houston 31
  - 2.3 Conclusions 35
- 3 Meteorological Clusters and Ozone 35
  - 3.1 Cluster Analysis 37
  - 3.2 Nonparametric Regression 38
  - 3.3 Urban Ozone in Houston 40
- 4 Regional Variation in Ozone 44
  - 4.1 Rotated Principal Components 45
  - 4.2 Gulf Coast States 46
- 5 Summary 48
- 6 Future Directions 48
- References 49

4 Design of Air-Quality Monitoring Networks   51
Douglas Nychka, Nancy Saltzman

- 1 Introduction 51
  - 1.1 Environmental Issues 51
  - 1.2 Why Find Spatial Predictions for Ozone? 52
  - 1.3 Designs and Data Analysis 53
  - 1.4 Chapter Outline 54
- 2 Data 55
  - 2.1 Hourly Ozone and Related Daily Summaries 55
  - 2.2 Handling Missing Data 56
  - 2.3 Model Output 56
- 3 Spatial Models 57
  - 3.1 Random Fields 57
  - 3.2 Spatial Estimates 60
  - 3.3 Design Evaluation 61
- 4 Thinning a Small Urban Network 62
  - 4.1 Preliminary Results 63
  - 4.2 Designs from Subset Selection 63
  - 4.3 Results 65
- 5 Adding Rural Stations to Northern Illinois 67
  - 5.1 Space-Filling Designs 68
  - 5.2 Results for Rural Illinois 71

	6	Modifying Regional Networks	71
		6.1 Results for the Larger Midwest Network	72
	7	Scientific Contributions and Discussion	74
		7.1 Future Directions	74
	References		75

5 Estimating Trends in the Atmospheric Deposition of Pollutants   77
David Holland

1	Introduction	77
2	Monitoring Data	78
	2.1 Case Study I	78
	2.2 Case Study II	78
	2.3 Additional Ongoing Monitoring	79
3	Case Studies	80
	3.1 Gamma Model for Trend Estimation	80
	3.2 Network Ability to Detect and Quantify Trends	82
4	Future Research	87
Acknowledgment		88
References		88

6 Airborne Particles and Mortality   91
Richard L. Smith, Jerry M. Davis, Paul Speckman

1	Introduction	91
2	Statistical Studies of Particles and Mortality	93
3	An Example: Data from Birmingham, Alabama	94
	3.1 Summary of Available Data	94
	3.2 Statistical Modeling Strategy	95
4	Results for Birmingham	102
	4.1 Linear Least Squares and Poisson Regression	102
	4.2 Nonlinear Effects	104
	4.3 Nonparametric Regression	108
5	Comparisons with Other Cities	111
	5.1 Seasonal Parametric and Semiparametric Models	112
	5.2 Results: Chicago	112
	5.3 Results: Salt Lake County	113
	5.4 Direct Comparisons Between Chicago and Birmingham	114
6	Conclusions: Accidental Association or Causal Connection?	118
References		119

7 Categorical Exposure-Response Regression Analysis of Toxicology Experiments   121
Minge Xie, Douglas Simpson

1	Introduction	121

		1.1	Critical Exposure-Response Information and Modeling Approaches 122

        1.2   Issues in Exposure-Response Risk Assessment 123
   2   The Tetrachloroethylene Database 124
        2.1   Severity Scoring 124
        2.2   Censoring 125
   3   Statistical Models for Exposure-Response Relationships 126
        3.1   Haber's Law 126
        3.2   Homogeneous Logistic Model 127
        3.3   Stratified Regression Model 129
        3.4   Marginal Modeling Approach 131
        3.5   Other Issues 132
   4   Computing Software: *CatReg* 134
   5   Application to Tetrachloroethylene Data 134
   6   Conclusions 136
   7   Future Directions 138
   Acknowledgments 138
   References 138

8 Workshop: Statistical Methods for Combining Environmental Information   143
Lawrence H. Cox

   1   The NISS-USEPA Workshop Series 143
   2   Combining Environmental Information 144
   3   Combining Environmental Epidemiology Information 145
        3.1   Passive Smoking 146
        3.2   Nitrogen Dioxide Exposure 148
   4   Combining Environmental Assessment Information 150
        4.1   A Benthic Index for the Chesapeake Bay 150
        4.2   Hazardous Waste Site Characterization 152
        4.3   Estimating Snow Water Equivalent 152
   5   Combining Environmental Monitoring Data 153
        5.1   Combining P-Samples 153
        5.2   Combining P- and NP-Samples 155
        5.3   Combining NP-Samples 155
        5.4   Combining NP-Samples Exhibiting More Than Purposive Structure 155
   6   Future Directions 156
   References 157

A   Appendix A: FUNFITS, Data Analysis and Statistical Tools for Estimating Functions   159
Douglas Nychka, Perry D. Haaland, Michael A. O'Connell, Stephen Ellner

   1   Introduction 159

	2	What's So Special About FUNFITS?	160
		2.1 An Example	161
	3	A Basic Model for Regression	165
	4	Thin-Plate Splines: tps	166
		4.1 Determining the Smoothing Parameter	167
		4.2 Approximate Splines for Large Data Sets	168
		4.3 Standard Errors	169
	5	Spatial Process Models: krig	169
		5.1 Specifying the Covariance Function	170
		5.2 Some Examples of Spatial Process Estimates	172
	Acknowledgments		178
	References		179

B Appendix B: DI, A Design Interface for Constructing and Analyzing Spatial Designs   181

Nancy Saltzman, Douglas Nychka

	1	Introduction	181
	2	An Example	182
	3	How DI Works	182
		3.1 Network Objects	182
		3.2 The Design Editor	182
		3.3 User Modifications	183

C Appendix C: Workshops Sponsored Through the EPA/NISS Cooperative Agreement   187

D Appendix D: Participating Scientists in the Cooperative Agreement   189

Index   191

Introduction: Problems in Environmental Monitoring and Assessment

Lawrence H. Cox
U.S. Environmental Protection Agency

Douglas Nychka
North Carolina State University

Walter W. Piegorsch
University of South Carolina

1 Statistical Methods for Environmental Monitoring and Assessment

The need for innovative statistical methods for modern environmental assessment is undisputed. The case studies in this book are a sampling of the broad sweep of statistical applications available in the environmental sciences, targeted to environmental monitoring and assessment. A unique feature of the applications presented here is that they are not isolated projects but were, instead, fostered under a long-term collaborative association between the U.S. Environmental Protection Agency (EPA) and the National Institute of Statistical Sciences (NISS). This institutional support resulted in a strong interdisciplinary component to the research, and common threads of statistical methodology and data analysis principles are seen across all of the projects. The case studies necessarily are detailed and technical and so this introductory chapter will give an overview of what follows and emphasize common themes that tie the projects together. Research, by its very nature, does not follow a direct path and depends on past results for the next step. This process is enriched through the collaboration of statisticians with other scientists.

A distinctive feature of the cooperative agreement was regularly scheduled research team meetings with EPA scientists and other researchers for each of the collaborative projects. For the modeling and design problems connected with ozone (see below), this was substantial, involving monthly discussions over the course of several years. Typically, a meeting would begin with a presentation of recent results, followed by discussion and suggestions for future directions. Close contact with EPA scientists helped to keep the projects relevant on the real environmental issues.

2 Outline of Case Studies

All of the projects making up the EPA/NISS collaboration have their roots in practical issues concerning the environment. Addressing these topics will often involve the use of advanced statistical techniques or the development of new methods. However, these studies are successful precisely because they answer practical questions posed by the environmental problems.

The first two chapters study surface-level ozone, a pollutant that has adverse health effects on humans and causes damage to crops and vegetation. Although it is important to quantify the extent of ozone pollution at different locations and its trend over time, ozone has variability at many different temporal and spatial scales which must be incorporated into any physically meaningful model. The research in these chapters ranges from trend quantification in ozone for the Chicago urban area to understanding the regional behavior along the Gulf Coast states. Meteorological conditions such as temperature and cloud cover control the formation of ozone, and through this work, strategies are given for using large numbers of meteorological variables in the analysis.

Chapter 4 on **network design** takes a more detailed view of the statistical issues involved in characterizing and modifying an air-quality monitoring network. Similar to Chapter 2, general methodology is studied that has applicability beyond the specific examples of network design. It is recognized, however, that broad-gauge research on statistical design often suffers from being too generic and lacks specific illustration. For this reason, Chapter 3 uses the current National Air Monitoring System/State and Local Monitoring Systems network as a test bed from which to propose specific network design modifications for predicting the spatial distribution of ozone. The results provide useful guidance on how to design, modify, and implement network systems for future use in atmospheric and environmental monitoring.

A related topic in air quality is the national effort to study and predict trends in the **atmospheric deposition** of gases and other atmospheric pollutants. The research effort involved modeling and predicting the complex factors affecting atmospheric deposition of aerosols on human populations, agriculture, and natural ecosystems. Chapter 5 gives an overview of the important results achieved in this endeavor, highlighting application of modern statistical methods such as gamma-GLM regressions and other regression models to estimate trend. Coupled with approaches that incorporate data into the decision process dynamically, these methods lead to recommendations for addition and/or deletion of monitoring network sites, similar to the goals discussed in Chapter 4, and/or covariates to better predict future changes in acid deposition.

Do high levels of **airborne particles** lead to increased rates of mortality? Recent attention has been given to an empirical link between particulate measurements (such as on matter of aerodynamic diameter less than 10 micrometers) made in urban areas and corresponding rises in mortality. Chapter 6 joins the debate through a careful examination of the regression models that are used in the analysis and illustrate the very real difficulties in drawing definitive scientific conclusions from observational studies.

Chapter 7 of the volume describes selected approaches used to model the toxic response in biological organisms after exposure to toxic stimuli. While a broad literature exists on this general issue of **environmental toxicology**, specifics for certain kinds of complex response endpoints have been lacking. This chapter addresses the important issue of modeling and analysis for ordered categorical outcomes, and describes modern, computer-intensive statistical methods appropriate for toxicological data. In addition, a customized computer software program is introduced that can facilitate the data analysis. As in previous chapters, the methods are discussed within the context of their future use with complex environmental data.

An important component of the EPA cooperative agreement was the sponsorship of workshops focused on important topics in environmental statistics. Chapter 8 focuses on the applications from the 1993 EPA/NISS workshop on **combining environmental information**. The effort of combining information often involves meta-analysis or similar forms of combination of statistical results, and there are many unresolved quantitative issues associated with such undertakings. For example, how do we combine results from several independent investigations about environmental endpoints of interest, each with a companion statistical analysis, into a single coherent conclusion? Several case studies, taken from the workshop, are used to illustrate standard and developing methodologies toward this goal. The contribution of this chapter extends beyond these illustrations, however, to the general issues of how to combine environmental information in different contexts.

3 Sources of Data and Software

The raw data used in these case studies were rarely in a form where it could be used directly. For instance, combining meteorological observations along with aerometric measurements (such as surface-level ozone) requires access to several databases. Also, in some instances, missing data had to be filled in to increase the number of available records. Understanding the details of these efforts is important to develop an appreciation for the realities of applied statistical projects, and to guide others in reproducing results. For these reasons, the case studies have separate sections that give sources of data and information as to how the raw data were handled.

Although it is important to be familiar with the basic data sources, it is also helpful to have available the specific data sets used in these cases studies. In this way, one can compare the results with other analyses or undertake related problems. Thus, to supplement these case studies, key data sets are available in electronic form and can be accessed from the NISS Web Site (http://www.niss.org/). Throughout this volume, this electronic, on-line resource will be referred to as the *web companion*. It is organized around the Table of Contents and will serve as a method for updating the case studies and providing links to authors and other researchers. Being able to reproduce work analyses and extend the results in this volume will enrich the case studies and serve as benchmarks for the development of new methods.

A crucial activity that supported this work was the development of statistical software for implementing the new methods. The volume concludes with a pair of appendices that describe computer software for fitting curves and surfaces (FUNFITS, in Appendix A) and for interactive construction and evaluation of spatial designs (DI, in Appendix B). The programs have bearing in many of the environmetric applications discussed above, and also in statistical activities covering a broad variety of other subject-matter areas. The software is publicly available and more information is available in the *web companion*.

Acknowledgments

We wish to acknowledge the kind support and useful comments of various colleagues, including the NISS Director, Jerome Sacks, and Associate Director, Alan F. Karr, and also R. Todd Ogden. Also, we wish to thank Elizabeth Rothney, Sharon Sullivan and Wanzhu Tu for help with manuscript preparation. The research described herein has been funded wholly or in part by the United States Environmental Protection Agency under assistance agreement #CR819638-01-0 to the National Institute of Statistical Sciences. Chapters authored by EPA (or NOAA) employees have been subjected to Agency (or NOAA) peer review, but remaining chapters and other portions of the book have not. The material contained herein represents the views of the individual authors and, therefore, does not necessarily reflect the views of the Agency. No official endorsement or Agency position should be inferred.

Modeling Ozone in the Chicago Urban Area

Jerry M. Davis
North Carolina State University

Brian K. Eder
National Oceanic and Atmospheric Administration[1]

Peter Bloomfield
North Carolina State University

1 Introduction

Ozone (O_3) is a ubiquitous trace gas in the atmosphere. Its highest concentration is in the stratosphere, where it shields the earth's surface from harmful ultraviolet radiation. At the surface, however, ozone is itself harmful, with destructive impacts on materials, crops, and health. Its levels have been high enough in certain areas to be of concern for several decades.

Ozone levels are difficult to control, as it is a *secondary pollutant*: It is not emitted directly into the atmosphere, but rather results from photochemical reactions involving *precursor* pollutants. The precursors include a variety of volatile organic compounds, comprised mainly of nonmethane hydrocarbons, nitric oxide (NO) and nitrogen dioxide (NO_2). Both nonmethane hydrocarbons and nitrogen oxides are emitted from transportation and industrial processes. Volatile organic compounds are emitted from diverse sources such as automobiles, chemical manufacturers, dry cleaners and other facilities using chemical solvents.

The rates and completeness of the reactions that produce ozone, as well as its subsequent transport and deposition, are driven by meteorological conditions, principally:

- the availability of sunlight;
- temperature; and
- wind speed (ventilation and transport).

Efforts to reduce emissions of precursors may, therefore, not be rewarded by prompt reductions in ozone concentrations; a shift in the weather to patterns more conducive to ozone production may more than offset a reduction in precursors. If observed changes in ozone concentrations could be adjusted for changes in these meteorological factors, then reductions in precursors would

---

[1] On assignment to the U.S. Environmental Protection Agency

be more apparent. Thus, many of the statistical contributions in this chapter are focused on determining the relationship between ozone concentrations and meteorology. Before giving an outline of the case studies, it is appropriate to include some background on the effects of ozone and attempts to regulate its levels.

1.1 Health and Environmental Effects

Elevated concentrations of ozone have been associated with effects lasting from hours to days on the human respiratory system, including decreases in functional lung capacity and lung flow rate (Lippmann 1989). In addition, high levels of ozone are strongly suspected as the cause of certain chronic respiratory effects. The U.S. Environmental Protection Agency (EPA) estimated that in 1990 nearly 70 million people lived in metropolitan areas that exceeded the national air quality standard then in effect for ozone.

While characterized initially as an urban-scale pollutant, ozone has increasingly been recognized as a regional (Logan 1989) and even global-scale phenomenon (Liu *et al.* 1987). High concentrations are routinely observed over vast, nonurban areas of most industrialized countries, and the consequent forest retardation and crop injury are growing environmental concerns (Lefohn and Lucier 1991). The Office of Technology Assessment (OTA 1989), in a review of the effects of ozone on crops in the eastern United States, determined that ozone reduces yields of economically important crops between 1% and 20%, and that annual benefits of between $0.5B (billion) and $1.0B would be realized if rural ozone concentrations were reduced by 25%. Ozone also has been hypothesized to be partially responsible for the decline of several tree species in the eastern United States, southern Canada, and Europe (OTA 1989; McLaughlin 1985).

1.2 Background

The Clean Air Act of 1971 required the U.S. EPA to establish standards for various pollutants, known as the National Ambient Air Quality Standards (NAAQS), to guide in monitoring and controlling their ambient concentrations. The primary standard[2] for ozone from 1978 until 1997, which was in effect when these studies were carried out, was based on the daily maximum ozone measurement among the network of stations monitoring a given area. The regulations stipulated that this daily network maximum could exceed 0.12 parts per million (ppm) only three times in a three year period.[3] Thus, if four

---

[2] Primary standards are designed to protect human health, and secondary standards are designed to protect human welfare.

[3] This standard was originally phrased as one exceedance per year, but because of the large annual influence of meteorology on ozone concentrations, it was subsequently relaxed to consider the average number of exceedances over three years.

or more exceedances were recorded over a three year period, the area was in violation of the standard.

The 1978 NAAQS 1-hour primary standard was replaced in 1997 by an 8-hour standard at a level of 0.08 ppm, in a form based on the 3-year average of the annual fourth highest 8-hour average ozone concentration measured at each monitor within an area. Thus, the new 8-hour, 0.08 ppm primary standard will be met at a monitoring site when the 3-year average of the annual fourth highest 8-hour average ozone concentration is less than or equal to 0.08 ppm.

Despite more than two decades of major control efforts, the ambient air concentration of ozone in both urban and rural areas remains a pervasive and persistent environmental problem (NRC 1991). To date, most regulatory efforts designed to attain the air-quality standards for ozone have not succeeded. This lack of progress in ozone abatement has disappointed regulators and perplexed air-quality scientists. Measuring this trend and interpreting it in light of the complicated relationship between meteorology and ozone formation was the initial goal of the research undertaken at the National Institute of Statistical Sciences (NISS).

1.3 Overview of the Case Studies

This chapter and the next describe several efforts to understand the impact of meteorology on ozone and to use this understanding in adjusting trend estimates. The first collection of studies is focused on the Chicago area.[4] The second collection of studies is focused on the southern United States, particularly Houston, an area with quite different meteorological patterns from Chicago.

A number of methods were used to develop trend estimates for the Chicago area. A nonlinear parametric model is discussed in Section 3; a nonparametric approach was adopted in the work covered in Section 4. Section 5 describes an approach based on modeling the frequency and magnitude of exceedances over a given level. The conclusions of these three analyses differ in detail, but broadly agree that while measured ozone concentrations have tended upward during the period covered (1981–1991), the adjusted trend is, in fact, negative, although not statistically significant.

For the analysis of urban ozone in Chicago and Houston and regional ozone in the Gulf Coast area, a number of advanced statistical procedures have proven useful. In Chicago, extensive use was made of nonlinear regression procedures, whereas in Houston, singular value decomposition coupled with cluster analysis was used to identify synoptic-scale scenarios (i.e., clusters of like meteorological days). Nonparametric regression techniques were then used to model ozone within each cluster for Houston. In this chapter and the next, each technique is discussed when it is first encountered in the analyses. It

---

[4]This region actually includes parts of Indiana and outlying suburbs (Figure 1).

will be assumed that the reader is generally familiar with linear regression analysis (Draper and Smith 1981); thus, the discussion will concentrate on some of the less familiar techniques and more modern statistical procedures. Many of the computations were carried out using the statistical package S-PLUS™ (Becker *et al.* 1988; Chambers and Hastie 1993); others were made using special-purpose programs developed by the various investigators.

The reader is referred to the *web companion* for specific data sets and software that are related to the case studies in this chapter.

2 Data Sources

2.1 Ozone Data

Ambient air measurements are collected by state and local agencies at nearly 800 stations nationwide. These agencies maintain standard operating procedures for air-quality monitoring in accordance with National Air Monitoring Systems/State and Local Air Monitoring Systems, collectively known as the NAMS/SLAMS network. The siting of stations is based on several criteria and is discussed in Chapter 4, but one feature relevant to this chapter is the characterization of stations as located either in a populated area or in a rural location. Ozone monitoring occurs during the "ozone season" (typically encompassing the 7-month period between 1 April and 31 October) and adheres to strict criteria established by the U.S. Environmental Protection Agency (1985) including calibrations, independent audits, and data validation, as well as a rigorous quality assurance program. Accepted ozone measurements are stored in EPA's Aerometric Information Retrieval System (AIRS) and are made available for public dissemination.

The instruments used are either chemiluminescence analyzers or ultraviolet photometers, and measurements can be taken continuously over time. However, the actual output is usually sampled at short time intervals (e.g., every minute) and these discrete measurements are then electronically averaged to produce a 1-hour average summary. Based on this preprocessing, we assume that the 1-hour ozone measurement, available in public databases, approximates an integrated average over the 1-hour period for which the measurement is reported.

2.2 Meteorological Data

The meteorological data used in these studies were obtained from the U.S. National Climatic Data Center and included both surface and upper-air measurements. Surface observations varied from study to study but typically included the components in Table 1.

The surface observations are available on an hourly basis; however, in the statistical studies described below, a daily summary value was often found to

atmospheric pressure: the force per unit area exerted on a surface by the random motion of air molecules. Pressure is proportional to the weight of an overlying air column of unit cross-sectional area (mb).

ceiling height: the height ascribed to the lowest layer of clouds when it is reported broken or overcast. (m)

dew-point temperature: the temperature to which air must be cooled (at constant pressure and water vapor) for saturation to occur. (°C)

mixing height: the height above the surface to which the atmosphere is well mixed. (m)

opaque cloud cover: the amount of sky cover that completely hides all that might be above it. (%)

relative humidity: the ratio of the amount of water vapor in the air compared to the amount required for saturation (at a given temperature and pressure). (%)

specific humidity: the ratio of the mass of water vapor in the air to the total mass of air. (g kg^{-1})

temperature: the degree of hotness or coldness of a substance as measured by a thermometer; alternatively, a measure of the average speed or kinetic energy of the atoms and molecules in a substance. (°C)

total cloud cover: the total amount of sky covered by all cloud layers. (%)

visibility: the greatest horizontal distance an observer can see and identify prominent objects. (km)

wind: the horizontal movement of air, with u being the west-to-east component and v the south-to-north component. Total horizontal wind speed is given by $\sqrt{u^2 + v^2}$. (m s^{-1}).

TABLE 1. Listing and some definitions of meteorological variables used in regression modeling of ozone.

be adequate.

Because the vertical structure of an air mass has a profound influence on the behavior of ozone (Eder et al. 1994), upper-air data were also used in some of the analyses. Upper-air observations are taken twice daily, at 0000 and 1200 UTC (Coordinated Universal Time, the new designation for Greenwich Mean Time, GMT), and for these studies, typically included temperature, dew-point depression and the u and v components of the wind recorded at the 850 mb and 700 mb pressure levels. In addition to the measured variables, several estimated variables deemed important to the study of ozone were also used, including daily incoming solar radiation ($MJ\,m^{-2}$) measured at the earth's surface, and the morning and afternoon mixing heights (Stull 1988).

3 Trend Analysis and Adjustment

A direct way to estimate the trend in a set of observed data, allowing for the effects of other variables such as meteorology, is to build a statistical model relating the response of the variable of interest to the other variables and to time. Linear models were found to be inadequate in modeling the complex relationship between ozone and meteorology. Linear and nonlinear models and their estimation are reviewed in the following section.

In prior work, Cox and Chu (1993) used generalized linear models to describe the relationship of surface ozone concentration measurements to meteorological variables in 43 urban areas in the United States, and obtained adjusted trend estimates for those areas. Section 3.2 discusses a NISS study of surface ozone data from the Chicago urban area (Bloomfield et al. 1996). A variety of methods, including fitting linear models and nonparametric models, were used to explore the dependence of the ozone concentrations on meteorology, and a detailed nonlinear parametric model was constructed and fit to the data.

3.1 Parametric Modeling

A regression function describes the relationship between a single covariate (X) or multiple covariates and the mean of a response variable Y. A basic linear model is given by

$$Y = \beta_0 + \beta_1 X_1 + \beta_2 X_2 + \ldots + \beta_p X_p + \varepsilon$$

where ε is the random error term, $E(\varepsilon) = 0$ and $\text{Var}(\varepsilon) = \sigma^2$. The parameters are typically estimated by least squares; the reader is referred to Draper and Smith (1981) for a comprehensive discussion of (linear) regression analysis.

In many analyses of environmental data, a nonlinear model relating a response to covariates is more appropriate, written

$$Y = f(\boldsymbol{X}, \boldsymbol{\beta}) + \varepsilon$$

where $\boldsymbol{X}$ is a vector of predictor variables and $\boldsymbol{\beta}$ is a vector of unknown parameters. For the nonlinear model, the error sum of squares is

$$S(\boldsymbol{\beta}) = \sum_{i=1}^{n}(Y_i - f(\mathbf{X}_i, \boldsymbol{\beta}))^2$$

and in the normal sampling case, the least squares estimate is also the maximum likelihood estimate. Many statistical systems including S-PLUS provide functions for specifying and fitting nonlinear models. For a small number of parameters, one can often use algorithms that numerically determine the gradient of $S(\boldsymbol{\beta})$, and thus require minimal amounts of user input. However, for larger and more complex nonlinear functions, explicitly specifying the partial derivatives of f with respect to $\boldsymbol{\beta}$ and carefully choosing initial values for the parameters are often necessary. For further discussion of nonlinear modeling, see Bates and Watts (1988) and Seber and Wild (1989).

3.2 Urban Ozone in Chicago

A Network Summary and Exploratory Analysis

The ozone data consisted of hourly averages at 45 stations in the Chicago area. Most of these stations recorded data only during the summer months, although some were operated essentially year-round. In all the analyses described subsequently, data were limited to the ozone season of 1 April to 31 October for the years 1981 to 1991. Surface meteorological data came from O'Hare Airport in Chicago, while the upper-air meteorological data came from the sounding facility at Peoria, IL (Figure 1).

To derive a simple representative ozone level for each day and to adjust for missing measurements, an additive decomposition was used. Let $y_{d,s}$ denote the maximum hourly concentration on day d at station s and let

$$y_{d,s} = \mu + \alpha_d + \beta_s + \varepsilon_{d,s}.$$

The parameter β_s is the level adjustment due to the station effect, and the adjustment due to day is given by α_d. The decomposition was effected by *median polish* (Tukey 1977), and the *daily network typical value* was then defined as $\mu + \alpha_d$. A similar strategy was used to fill in missing data at the station level, and in this case, the additive effects are hour and day. The median polish technique is related to least absolute deviations methods and is not sensitive to extreme values. For this reason, it was preferred to estimating $\mu + \alpha_d$ using least squares. Although a precise definition of the median polish decomposition is related to the computational algorithm, roughly speaking one could interpret the network typical value as an estimate of the network *median*.

The relationship of the network typical value to meteorological variables was explored using graphs and nonparametric fitting methods such as *loess* (see Chapter 3, Section 3.2). For example, Figure 2 displays the dependence of ozone on maximum temperature and wind speed. Here, we see a complicated relationship of ozone with wind speed, approximately linear for low temperatures but undergoing a transition to a nonlinear shape for high temperatures.

A nonlinear model relating the network typical ozone value to meteorological covariates was then constructed in an iterative fashion. Beginning with a function designed to mimic the dependence shown in Figure 2, residuals from each

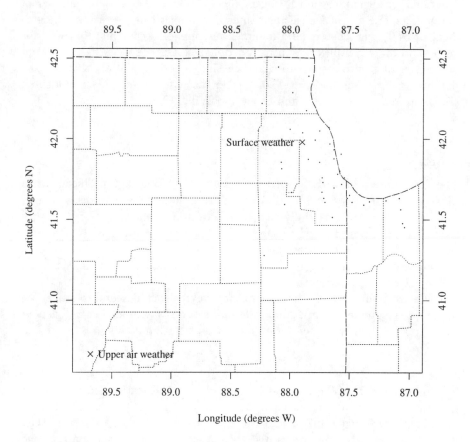

FIGURE 1. Locations of ozone monitoring stations (dots) and weather stations.

tentative model were related to other variables, and the model was extended and modified to include each variable that seemed to be important. Parameter estimates were reviewed at each step to identify possibilities for simplifying the model. The process led to a nonlinear model, relating the network typical ozone value with the meteorological covariates in Table 1.

If the model was intended to be used primarily for forecasting ozone concentrations, its predictive power could be improved by including the previous days' ozone levels as additional covariates. However, it would then be an equation for the *conditional* mean of ozone given past ozone levels (*inter alia*), whereas any trend is part of the *unconditional* mean. Converting one equation to the other is straightforward for linear models but far less so for nonlinear models, and for this reason, lagged ozone levels were not used. Lagged values for a number of the meteorological covariates, however, were found to be necessary. In addition, residuals from the tested models exhibited a seasonal dependence

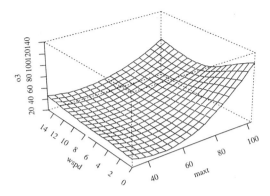

FIGURE 2. Loess estimate of ozone as a function of temperature and wind speed.

which necessitated the addition of a seasonal term. Finally, a trend term was included in the model.

The Model Equation

The final form of the model with meteorological variables is given below. The dependent variable, O_3, is the daily network typical ozone. Coefficients which are fit to the model are italicized in the model equation and Roman abbreviations identify the independent variable.[5] The equation is

$$O_3 = (\mu_0 + (t_0 + t_1 \text{maxt} + t_2 \text{maxt}^2 + t_3 \text{maxt}^3 + t_{l1}\text{tlag1} + t_{l2}\text{tlag2})$$

$$\times 1/(1 + \text{wspd}/v + \text{wspd700}/v_{700} + \text{wlag}/v_l)$$

$$\times (1 + r\,\text{rh} + r_l\text{rhlag})(1 + o\,\text{opcov})(1 + \omega\,\text{vis})$$

$$\times (1 + m_u\text{mean.u} + m_v\text{mean.v})(1 + \tau\,\text{year}))$$

$$+ (a_1\,\text{c1} + b_1\,\text{s1} + a_2\,\text{c2} + b_2\,\text{s2})$$

Here, maxt is the daily maximum temperature, tlag is its lagged value, wspd is noon surface wind speed, wspd700 is the 12 UTC 700 mb wind speed, rh is the noon relative humidity, rhlag is the lagged value, opcov is the noon opaque cloud cover, vis is the noon visibility, mean.u is the 24-hour average

---

[5]The meteorological variables have been centered around the following values: maxt 60°F, tlag 60°F, rh 50%, rhlag 50%, opcov 50%, and vis 12 km. The centering helps to improve interpretability of the model parameters. Although the notation for this model may seem awkward, it facilitates identifying the parameter estimates in Bloomfield et al., (1996).

west-to-east surface wind component, mean.v is the 24-hour average south-to-north surface wind component, year is the time in units of years, and the final terms are annual and semiannual cosine and sine terms:

$$c1 = \cos(2\pi \text{year}) - \text{ave}\{\cos(2\pi \text{year})\}$$
$$s1 = \sin(2\pi \text{year}) - \text{ave}\{\sin(2\pi \text{year})\}$$
$$c2 = \cos(4\pi \text{year}) - \text{ave}\{\cos(4\pi \text{year})\}$$
$$s2 = \sin(4\pi \text{year}) - \text{ave}\{\sin(4\pi \text{year})\}$$

and so on.

The multiplicative form of the mean model (other than the seasonal terms) was found to fit better than an additive version and implies that the components contribute to relative changes in the mean level of ozone.[6] The leading "1" in all but the first component makes all parameters identifiable. Note that because the trend in this model, τ, enters as a multiplicative term, it may be interpreted as the relative change with respect to the mean level. This meshes well with the common form of reporting environmental trends as a percentage change per decade. The reader should keep in mind that this final model was arrived at after extensive data and residual analysis, and the nonlinear form for the temperature and wind speed was the result of examining several nonparametric models and plotting residuals against these meteorological variables.

Predictions from this model were within ±5 ppb about half the time, and within ±16 ppb about 95% of the time. The root mean square residual was 8.2 ppb; however, residual analysis indicated that the residual component was not normally distributed. Model consistency was tested using cross-validation techniques and confidence intervals for the parameter estimates were arrived at using a jackknife procedure.[7] The jackknife procedure was warranted due to the nonlinear form of the model and due to the non-normal distribution of the random component.

Conclusions

The network typical value time series showed nonlinear and nonadditive dependence on various meteorological quantities, including individual measurements and averaged values. There was also strong seasonal dependence, even after allowing for the effects of the meteorological variables. This dependence was approximated by a nonlinear parametric model, and when the parameters were fitted by least squares, the model accounted for 80% of the variance of the ozone concentration data. A model that included only seasonal and trend

---

[6] Strictly, the components contribute to relative changes in the *multiplicative part* of the mean function. Since the seasonal terms were centered over the entire data set and hence almost centered in each year, they leave this interpretation approximately intact.

[7] Jackknifing was implemented by leaving out year-long blocks of data.

components explained only 30% of the variation in the daily network typical ozone; thus, meteorology was important. The model was able to predict moderately high ozone values quite well; however, for extreme ozone values, the model underestimated the observed ozone.

Trends in ozone with and without meteorology were evaluated. Neither of the trend estimates was significantly different from zero, but adjustment for meteorology yielded a narrower confidence interval. However, even the best model yielded a wide 95% confidence interval (−10.3%, 4.9%) for trend per decade. The standard error of the trend coefficient, adjusted for heteroscedasticity and short-term serial correlation (Gallant 1987, Sections 2.1 and 2.2), was found to be 1.4% per decade. This increased to 3.4% per decade when a jackknife procedure was used, apparently because of small correlations on annual and longer time scales. It would appear that even substantial linear trends in ozone concentrations may not be detected for some years.

3.3 Comparison with Rural Location

As a first attempt to determine whether the insights yielded by the study of urban ozone variability in Chicago were applicable elsewhere, Bloomfield et al. (1993) studied ozone concentrations and meteorology in the rural area surrounding Chicago. The model described in Section 3.2 was refitted to data from the rural network and was found to give a comparably good fit. In a further study, Steinberg and Bloomfield (1994) reexamined the data from the rural network. Since some of the rural stations were located closer to Detroit or St. Louis than they were to Chicago, Steinberg and Bloomfield incorporated the meteorology from these locations into the rural model. The results indicated that model performance was not improved by the additional meteorological data. The authors suggested that the rural ozone stations did not form a homogeneous network with respect to meteorological influences, and that the meteorological data for Chicago, Detroit, and St. Louis be used for three geographically determined subnetworks of stations. The model was then found to fit reasonably well for each subnetwork.

The resulting meteorologically adjusted trend estimates for the Chicago, St. Louis, and Detroit subnetworks were −2.0% (5.6%) per decade, −12.5% (16.8%) per decade, and +0.5% (0.4%) per decade, respectively. The authors found that the trend parameters differed significantly only for the St. Louis/Detroit comparison. The implication is that the meteorologically adjusted ozone trends differ between these two subnetworks; but, no conclusion could be drawn from the analysis regarding the Chicago/St. Louis or Chicago/Detroit pairs.

4 Trends from Semiparametric Models

The model carefully constructed in Section 3 represents the apparent relationship of ozone concentrations with each of the meteorological variables. However, it was acknowledged to be an approximate, empirically obtained model, and fitting a misspecified model may produce biased estimates, most importantly, of the trend. An alternative approach is to fit an arbitrary smooth function of the meteorological variables. Such an approach eliminates the structural assumptions embodied in a parametric model, replacing them with qualitative assumptions such as smoothness, and allowing exploration of the sensitivity of trend estimates to such structural assumptions.

Some of the many approaches to smooth function estimation are discussed in the following chapter. One, based on *semiparametric* modeling, is the subject of the next section and is applied to the Chicago urban ozone data in Section 4.2.

4.1 Semiparametric Modeling

The modeling procedure discussed below[8] was developed by Sacks et al. (1989) in prior work on the analysis of computer experiments involving no measurement error. Its first use in the presence of measurement error was in a NISS study described by Gao et al. (1996). This approach has some striking similarities to spatial prediction and Kriging, but it is not restricted to spatial observations. Also, the covariance function has parameters estimated by maximum likelihood.[9]

The observational model has the form

$$Y(\mathbf{x}) = \sum_{j=1}^{k} \beta_j f_j(\mathbf{x}) + Z(\mathbf{x}) + \varepsilon,$$

where $\mathbf{x} = (x_1, \ldots, x_p)$ consists of the p covariates in the model. The β_j are parameters associated with *fixed* functions f_j, and $\varepsilon \sim N(0, \sigma_\varepsilon^2 I)$. We assume $Z(\mathbf{x})$ is a zero mean Gaussian process with covariance function

$$\mathrm{cov}(Z(\mathbf{x}), Z(\mathbf{x}')) = \sigma_Z^2 \exp\left(-\sum_{i=1}^{p} \theta_i |x_i - x_i'|^{p_i}\right),$$

and $1 \leq p_i \leq 2$.

The smoothness of the response surface as a function of the covariates is indicated by p_i ($p_i = 2$ corresponds to a high level of smoothness). Relative importance of variables is indicated by their corresponding values of θ when the variables are on normalized scales, with larger values of θ indicating greater importance. With this model, those coefficients θ which are zero correspond

---

[8] Also known as a Gaussian Additive Spatial Process, GaSP.
[9] See Chapter 3, Section 4 and Appendix A for more discussion on spatial models.

to covariates not selected for use; the others are selected. In the context of spatial statistics and Kriging, one can identify σ_ε^2 as the nugget variance, σ_Z^2 as the sill, the θ_i as range parameters, and, finally, the p_i as determinators of the smoothness of the function. The parameters σ_Z^2, σ_ε^2, θ, and p are estimated by maximum likelihood.

Given data $(\mathbf{x}_1, y_1), (\mathbf{x}_2, y_2), \ldots, (\mathbf{x}_n, y_n)$ and if σ_Z, σ_ε, and $R(\cdot, \cdot)$ are known, the best linear unbiased predictor, $\hat{y}(\mathbf{x})$, at a new point $\mathbf{x}$ is

$$\hat{y} = \sum_{j=1}^{k} \hat{\beta}_j f_j(\mathbf{x}) + \hat{Z}(\mathbf{x}) = \sum_{j=1}^{k} \hat{\beta}_j f_j(\mathbf{x}) + \mathbf{r}'(\mathbf{x}) C^{-1}(\mathbf{y} - F\hat{\boldsymbol{\beta}}),$$

where $\mathbf{y} = (y_1, y_2, \ldots, y_n)$, $C = \text{Corr}(y) = (\sigma_Z^2/\sigma^2)R + (\sigma_\varepsilon^2/\sigma^2)I$, and $\sigma^2 = \sigma_Z^2 + \sigma_\varepsilon^2$. $R = \{R(\mathbf{x}_i, \mathbf{x}_j), 1 \leq i \leq n; 1 \leq j \leq n\}$, the $n \times n$ matrix of correlations among values of Z at the data points, $\mathbf{r}(\mathbf{x}) = (\sigma_Z^2/\sigma^2)[R(\mathbf{x}_1, \mathbf{x}), \ldots, R(\mathbf{x}_n, \mathbf{x})]'$, $F_{jk} = f_k(\mathbf{x}_j)$, and $\hat{\boldsymbol{\beta}} = (\hat{\beta}_1, \ldots, \hat{\beta}_k)' = (F'C^{-1}F)^{-1}F'C^{-1}\mathbf{y}$, which is the usual generalized least squares estimate of $\boldsymbol{\beta} = (\beta_1, \ldots, \beta_k)'$.[10]

4.2 Application to Chicago Urban Ozone

Gao et al. (1996) applied the modeling technique described in the previous section to the same stations and time period as in Section 3. This method does not require a parametric model for the mean, and thus has fewer assumptions than those of Section 3. One disadvantage is that one may expect larger standard errors due to more flexibility in the model. The model used to predict future ozone levels based on observed meteorology had the basic form given in Section 4.1. Previous experience with computer experiments suggested that simple choices for the fixed part of the model are usually adequate; the Gaussian process $Z(\cdot)$ usually adapts to the more complex part. In the present case, year to year changes of average levels could be important, however, and so the fixed part was parameterized to capture level shifts for each year. Accordingly, let β_r be the mean level of ozone for year r. The relationship between these parameters and a trend will be detailed below.

The Gao et al. study focused on the summer period 15 May to 15 September (1981–1991) when ozone concentrations were relatively high. This four-month period was subdivided into four equal parts and each part was fit separately using the eleven years of data (see Table 2). Subdividing was important for two reasons: the assumption of stationarity within a shorter time period is more plausible than for a longer period, and the computational burden associated with the likelihood evaluation was reduced.

The authors fitted the model for the four time periods using network typical ozone as the response variable and various meteorological variables as the

---

[10]This covariance function is part of FUNFITS in Appendix A and the predictions and estimates of β (for given θ and p) can be computed using the function krig

Period	Trend	Std. Error	t Value
15 May–15 June	1.39	3.55	0.39
15 June–15 July	−6.35	3.11	−2.04
15 July–15 Aug.	−0.74	4.65	−0.16
15 Aug.–15 Sept.	−10.94	5.82	−1.88

TABLE 2. Trend estimates (% per decade) and jackknifed standard errors from spatial prediction model.

covariates. Temperature, relative humidity, and wind variables (surface wind speed and its one day lagged value, the mean u and v components of the surface wind, and the 700 mb wind speed) were important for all four time periods. The authors also explored main and joint effects of some of the important meteorological variables; their results were similar to those found in Section 3.

The quality of the model was assessed by calculating the cross-validation prediction at the ith data point x_i using a model fit with the remaining data. The cross-validation prediction was calculated at all data points and compared with the actual network typical value at that point. The prediction errors based on this cross-validation exercise were found to be low; however, the model underpredicted when the ozone levels were high. This problem was also encountered using the nonlinear model developed in Section 3.2.

The parameter β_r can be interpreted as the average ozone for the rth year having been adjusted for meteorological effects. To determine trends in ozone concentration levels after adjusting for meteorology, a linear trend was fit to the estimated year parameters β_r. Note that because the summer ozone season was divided into four parts, the analysis actually produced four trend estimates. Trend estimates were then converted into a "percent per decade" form:

$$\widehat{\text{trend}} = 100 \times \frac{(\hat{b}/10)}{\hat{a}} \quad (\%/\text{decade})$$

where $\hat{a}$ is the intercept at year $= 1981$ and $\hat{b}$ is the linear trend. In addition to standard errors using the stochastic model, a jackknifing procedure was also used to arrive at approximate confidence intervals for the parameter estimates and the results are shown in Table 2.

If the trend estimates reported here are averaged over the four periods, they are broadly consistent with the parametric analyses in Section 3. By dividing the ozone season into periods, however, the authors were able to detect significant trends in the network typical value within two of the four periods. More importantly, their analysis suggests that the trend in network maximum values reported in Section 3 is largely due to a reduction in two of the four periods.

The authors also tested the model as a predictive tool by developing the model over the years 1981–1987, and then predicting ozone levels in 1988 (a high ozone year) using the observed meteorology. The predictions in the four periods were generally good; however, the model had difficulty capturing the peaks in this severe year, as in other years.

5 Trends in Exceedances

Trends in ozone concentrations were estimated in Sections 3 and 4 by fitting models to the full distribution of ozone levels, as functions of meteorological variables. The focus of regulation, however, is on high levels, and especially on how often certain levels (such as 120 ppb) are exceeded. Trends in the mean may be of less interest than trends in the frequency of such exceedances. While the probability of exceeding a given level is determined by the mean and an assumption about the shape of the distribution, these probabilities may be estimated more directly by studying the occurrence of high concentrations in the observed data. Some statistical models for exceedances and extreme values are discussed in the next section. They are then used to study trends in high concentrations in the Chicago data in Section 5.2

5.1 Exceedance Modeling

Exceedance models refer to probability distributions that describe the frequency of extreme events. Smith and Huang (1993) point out that one disadvantage of regression analysis is its failure to account for what is happening in the extremes of the distribution. Given that both previous and current ozone regulatory standards are based on the rare occurrence of large ozone values, a regression model for the mean value of ozone may not be appropriate for predicting extreme events. Models based on regression have indeed tended to underpredict high ozone levels. This may result from the model-fitting procedures: minimizing the residual sum of squares over the whole data set with less direct attention to the extremes of the data.

Classical extreme value theory is that branch of statistics focusing on distributions of extreme values of random samples. Standard limiting distributions of extreme values follow one of three types of distributions (Leadbetter *et al.* 1983). In modern statistical applications, these are often combined into a single three-parameter *generalized extreme value* (GEV) family defined by the distribution function

$$H(y; \mu, \psi, \xi) = \exp\left[-\left\{1 + \xi \frac{(y-\mu)}{\psi}\right\}_+^{-1/\xi}\right]$$

(here $x_+ = \max(x, 0)$) where μ, ψ, and ξ represent, respectively a location, scale, and shape parameter of the distribution.

An extension of the extreme value approach is the "threshold" approach (Davison and Smith 1990), which turns out to be ideally suited to the present study. In this procedure, all ozone levels above a certain threshold are included in the analysis, and the problem of specifying a suitable distribution is applied to the *excesses* over that threshold. By analogy with the classical theory, there is also a class of limiting distributions for excesses, the family of *generalized Pareto distributions* (GPD). This is given by

$$G(x;\sigma,\xi) = 1 - \left(1 + \xi\frac{x}{\sigma}\right)_+^{-1/\xi},$$

where $x > 0$ is the excess over the threshold, and σ and ξ are scale and shape parameters to be estimated. These models can be extended so that the parameters σ and ξ depend on covariates; in this way, it is possible to incorporate meteorological dependence into the model. For some additional comments on the estimation of such "tail parameters," see Leadbetter (1993).

5.2 Modeling Exceedance Probabilities for the Chicago Urban Area

In Smith and Huang (1993), exceedance modeling was applied to a number of data sets extracted from the Chicago ozone study (Section 3). One goal of the research was to model the probability that ozone on a given day exceeded a specified threshold (120 or 150 ppb), as a function of a set of covariates. The authors selected three stations where ozone levels were high and also considered daily maxima across the network. Separate days were assumed to be independent, and with this assumption, the likelihood for the data is given by

$$L = \prod_i (p_i)^{\delta_i}(1-p_i)^{1-\delta_i},$$

where p_i gives the probability that the threshold is exceeded on day i, and δ_i is an indicator of whether the threshold is, in fact, exceeded on day i. A logit model was used for p_i,

$$\log(p_i/(1-p_i)) = \sum_j x_{ij}\beta_j,$$

where x_{ij} is the value of the jth covariate on day i and β_j is the corresponding coefficient. The model covariates were selected by a stepwise procedure where the minimized values of the log likelihood were used to decide which variables to introduce and when to stop.

Model results for the time period used in the study indicated a downward trend in the rate of ozone exceedances after accounting for meteorology. Fitting a linear trend for exceedances over 120 ppb gave statistically significant downward trends at two of the three individual stations and for the network maxima. Results for exceedances over 100 ppb were less significant. For one of the three stations and the network maxima, the authors found evidence for a quadratic trend. The fitted model suggested a slight increase in ozone levels up to 1984/1985, followed by a decrease in subsequent years.

The predictive ability of the model was also assessed. A year-by-year analysis based on the model with no trend showed a significant underprediction of exceedances in 1984 and 1985 and a significant overprediction in 1989 and 1991. The inclusion of a quadratic trend term and a previous day's exceedance term resulted in improved predictions. Model results were also compared with the nonlinear regression model of Section 3. For two of the three stations, the authors found that the nonlinear model offered an improvement, while for the third station and the network maxima it did not. Taking into account the added computational effort required to fit the nonlinear model, the extra complication does not seem to be justified.

5.3 Modeling Excess Ozone over a Threshold

To gain a real understanding of the extreme behavior of the ozone levels, however, one must go beyond modeling the exceedance probabilities of a fixed threshold. Smith and Huang also considered the excesses over a threshold, i.e., the amounts by which the threshold was exceeded on a given day. The generalized Pareto distribution (GPD) was used to model the excesses over the high thresholds. This model provided a good fit to the data.

A general framework for the GPD procedure was laid out by Davison and Smith (1990). They considered regression models in which the excess (if one occurs) on day i, say, is represented by the GPD cdf $G(\cdot; \sigma_i, \xi_i)$ with σ_i and ξ_i depending on covariates. In practice, it is usually adequate (and simpler) to assume ξ constant, while the interpretation of σ_i as a scale parameter suggests naturally that a logarithmic link function would be appropriate. One is led to consider models of the form

$$\log \sigma_i = \sum_j x_{ij} \gamma_j, \quad \xi_i = \xi,$$

in terms of new regression coefficients $\{\gamma_j, \ j = 1, 2, ...\}$. To give one example of the results of this analysis, Figure 3 shows probability plots of residuals (observed vs. expected values of the order statistics) under two different models for the distribution of excesses among the network maxima: (a) the exponential distribution and (b) the GPD. The exponential distribution is a limiting form of the generalized Pareto, corresponding to $\xi \to 0$. The residual plot shows that the exponential distribution is, in fact, rather a poor fit for this example, whereas the generalized Pareto appears adequate. This is an interesting conclusion because it is widely believed that exceedances in ozone series follow an exponential distribution. The current analyses have confirmed this when the ozone values are fitted *without* meteorological covariates. But in a regression

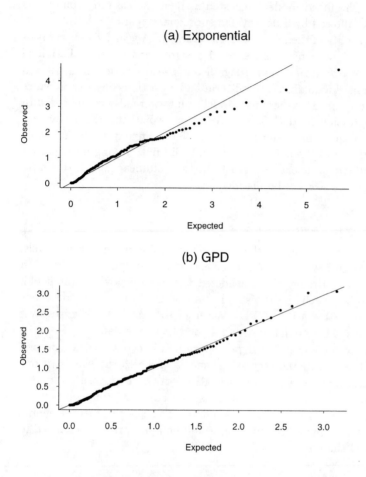

FIGURE 3. Probability plots of residuals (observed vs. expected values of the order statistics) under two different models for the distribution of excesses among the network maxima: (a) exponential distribution and (b) generalized Pareto.

model such as this, with adjustment made for meteorological conditions, the exponential assumption does not appear adequate.

An alternative approach based on Poisson process modeling (Smith 1989) was employed in analysis of data from Houston (Smith and Shively 1995). This again extended an earlier practice by allowing all the parameters of the exceedance model to depend in quite a general way on meteorological covariates. An advantage of the Poisson process approach is that there is just one set of parameters to be modeled, rather than modeling the probability of exceedance and the excess over the threshold by completely separate distributions. Unfortunately, when this approach was applied to the Chicago data, it was found to be less successful than the GPD approach, apparently because the GPD approach allowed greater flexibility in the final form of model adopted.

Given a model for threshold crossing probabilities and the distribution of excesses over a threshold, Smith and Huang showed how to compute crossing probabilities and the expected number of exceedances with respect to any level larger than the original threshold. Predictive assessments were carried out at four specific ozone levels. In 1988, the model for the network maxima (including trend) underpredicted the true rate of exceedances, and also to a lesser extent in 1984 and 1989.

5.4 Prediction of Extreme Ozone Levels

A predictive, cross-validation approach was developed as a way of assessing how well the model was actually performing at high threshold levels. For each of the 11 years of data, each year was successively omitted and the model refitted to the remaining 10 years (cross-validation). Then, the refitted model was used to "predict" daily ozone levels for the omitted year. For each of a number of high threshold levels, a predictive score was used to assess the agreement between forecast and observed probabilities of exceedance of the threshold, using methods adapted from Seillier-Moiseiwitsch and Dawid (1993). The results showed a consistent difficulty in forecasting the observed highest levels for 1988, and to a lesser extent in 1984 and 1989, based on the assumed model. This difficulty was ultimately resolved by removing one of the original assumptions of the analysis, namely that successive days are independent given the meteorological conditions. An alternative extremal model based on Markov dependence was employed (Smith 1994) and was shown to alleviate the earlier difficulty with predicting the most extreme levels.

The authors also examined 1988, which was a year of high ozone levels. Their results lend some support to the notion that 1988 was the most extreme year in terms of meteorology dating back to 1959. The model underpredicted the actual ozone levels that occurred in 1988. In this, it was consistent with the other two models fitted to the Chicago data; the parametric model described in Section 3 actually *over*predicted 1988 levels on average, but systematically underpredicted high observed levels in all years, as did the nonparametric model of Section 4. Thus, there is some support for the contention that unmodeled factors were responsible for the exceptionally high levels of that year.

6 Summary

The Chicago-area studies described in this chapter showed that statistical models guided by atmospheric science can explain much of the variability in surface ozone concentrations in terms of meteorological variability and seasonality. They also showed that the effects of meteorology need to be taken into account when measuring and interpreting ozone trends. The findings, although not always statistically significant, were that actual increases in ozone levels were associated with differences in meteorology, and it was estimated that ozone would, in fact, have decreased if meteorology could have been held constant.

The Chicago area is of interest in its own right as a major urban area and as a representative of similar interior cities. There was considerable interest in whether the strategies that are described above could be applied to other, less similar areas. The next set of studies moved on to a region with somewhat different characteristics, namely the Gulf Coast states. Chapter 3 discusses some of the results.

Acknowledgments

The authors appreciate the careful reading of the ozone chapters by Marlina Nasution. Although the data ozone used in this chapter are in the public domain, the project benefited greatly from the expert help of William Cox from the EPA for helping NISS researchers with the AIRS database and explaining the intricacies of these monitoring data.

References

Bates, D.M. and Watts, D.G. (1988). *Nonlinear Regression Analysis and its Applications.* Wiley, New York.

Becker, R.A., Chambers, J.M. and Wilks, A.R. (1988). *The New S Language: A Programming Environment for Data Analysis and Graphics.* Wadsworth and Brooks/Cole, Pacific Grove, CA.

Bloomfield, P., Royle, A. and Yang, Q. (1993). Rural ozone and meteorology: analysis and comparison with urban ozone. Technical Report 5. National Institute of Statistical Sciences, Research Triangle Park, NC.

Bloomfield, P., Royle, A., Steinberg, L.J. and Yang, Q. (1996). Accounting for meteorological effects in measuring urban ozone levels and trends. *Atmospheric Environment* **30**, 3067–3078.

Chambers, J.M. and Hastie, T.J. (eds.) (1993). *Statistical Models in S.* Chapman & Hall, New York.

Cox, W.M. and Chu, S.-H. (1993). Meteorologically adjusted ozone trends in urban areas: A probabilistic approach. *Atmospheric Environment* **27B**, 425–434.

Davison, A.C. and Smith, R.L. (1990). Models for exceedances over high thresholds (with discussion). *Journal of the Royal Statistical Society, series B* **52**, 393–442.

Draper, N. and Smith, H. (1981). *Applied Regression Analysis*. Wiley, New York.

Eder, B.K., Davis, J.M. and Bloomfield, P. (1994). An automated classification scheme designed to better elucidate the dependence of ozone on meteorology. *Journal of Applied Meteorology* **33**, 1182–1199.

Gallant, A.R. (1987). *Nonlinear Statistical Models*. Wiley, New York.

Gao, F., Sacks, J. and Welch, W.J. (1996). Predicting urban ozone levels and trends with semiparametric modeling. *Journal of Agricultural, Biological, and Environmental Statistics* **1**, 404–425.

Leadbetter, M.R. (1993). On high level exceedance modeling and tail inference. Technical Report 8. National Institute of Statistical Sciences, Research Triangle Park, NC.

Leadbetter, M.R., Lindgren, G. and Rootzen, H. (1983). *Extremes and Related Properties of Random Sequences and Processes*. Springer-Verlag, New York.

Lefohn, A.S. and Lucier, A.A. (1991). Spatial and temporal variability of ozone exposure in forested areas of the United States and Canada: 1978–1988. *Journal of the Air & Waste Management Association* **41**, 694–701.

Lippmann, M. (1989). Health effects of ozone: A critical review. *Journal of the Air & Waste Management Association* **39**, 672–695.

Liu, S.C., Trainer, M., Fehsenfeld, F.C., Parrish, D.D., Williams, E.J., Fahey, D.W., Hubler, G. and Murphy, P.C. (1987). Ozone production in the rural troposphere and the implications for regional and global ozone distribution. *Journal of Geophysical Research* **92**, 4191–4207.

Logan, J.A. (1989). Ozone in rural areas of the United States. *Journal of Geophysical Research* **94**, 8611–8532.

McLaughlin, S.B. (1985). Effects of air pollution on forest: A critical review. *Journal of the Air Pollution Control Association* **35**, 512–534.

National Research Council (1991). *Rethinking the Ozone Problem in Urban and Regional Air Pollution*. National Academy Press, Washington DC.

Office of Technology Assessment (1989). *Catching Our Breath: The Next Steps For Reducing Urban Ozone*. Office of Technology Assessment, Washington, DC.

Sacks, J., Welch, W.J., Mitchell, T.J. and Wynn, H.P. (1989). Design and analysis of computer experiments. *Statistical Science* **4**, 409–435.

Seber, G.A.F. and Wild, C.J. (1989). *Nonlinear Regression.* Wiley, New York.

Seillier-Moiseiwitsch, F. and Dawid, A.P. (1993). On testing the validity of sequential probability forecasts. *Journal of the American Statistical Association* **88**, 355–359.

Smith, R.L. (1989). Extreme value analysis of environmental time series: An application to trend detection in ground-level ozone (with discussion). *Statistical Science* **4**, 367–393.

Smith, R.L. (1994). Multivariate threshold methods. *Extreme Value Theory and Applications*, J. Galambos, J. Lechner and E. Simiu (eds.). Kluwer Academic Publishers, Dordrecht, pp. 225–248.

Smith, R.L. and Huang, L.-S. (1993). Modeling high threshold exceedances of urban ozone. Technical Report 6. National Institute of Statistical Sciences, Research Triangle Park, NC.

Smith, R.L. and Shively, T.S. (1995). A point process approach to modeling trends in tropospheric ozone. *Atmospheric Environment* **29**, 3489–3499.

Steinberg, L.J. and Bloomfield, P. (1994). Evaluation of data from three meteorologic stations for the parametric modeling of rural ozone. Technical Report 17. National Institute of Statistical Sciences, Research Triangle Park, NC.

Stull, R.B. (1988). *An Introduction to Boundary Layer Meteorology.* Kluwer Academic Publishers, Boston.

Tukey, J.W. (1977). *Exploratory Data Analysis.* Addison-Wesley, Reading, MA.

U.S. Environmental Protection Agency (1985). *Quality Assurance Requirements for State and Local Monitoring Stations (SLAMS).* 40 CFR, Part 58, Appendix A. Environmental Protection Agency, Washington, DC.

Regional and Temporal Models for Ozone Along the Gulf Coast

Jerry M. Davis
North Carolina State University

Brian K. Eder
National Oceanic and Atmospheric Administration[1]

Peter Bloomfield
North Carolina State University

1 Introduction

The studies described in the previous chapter focused on estimating trends in a daily ozone summary having adjusted for the relationship of surface ozone concentrations to meteorology. Moreover, the analysis was largely restricted to the Chicago urban area. This chapter contrasts this narrow scope by studies that:

- model the entire day's ozone profile rather than its maximum;

- relate ozone levels to specific weather patterns;

- characterize the ozone field at different spatial scales from urban to regional.

To broaden the results from the previous chapter, these investigations were carried out in the Gulf Coast states of the United States, an area that has a different climate from the Midwest.

1.1 Scientific Issues

The Chicago studies were based entirely on daily maximum observed concentrations. While the daily maximum is an important feature of the 24 hourly values, it is far from a complete summary of the day's ozone levels and should not be the exclusive focus of analysis. How much structure is there in the hourly ozone values over a day? It is clearly desirable to work with a low-dimensional representation to aid in interpreting diurnal patterns and also to reduce the size of the data sets. The first case study describes one approach (see Section 2).

---

[1] On assignment to the U.S. Environmental Protection Agency

The analysis of the Chicago area was successful in building complex regression functions that relate the mean ozone (or its extreme values) to local meteorological variables. Besides being effective for trend detection, the regression function can also suggest the meteorological conditions that are conducive for ozone formation. Thus, one scientific contribution of the second case study in this chapter is to associate high or low ozone levels with physically meaningful meteorological patterns. Another benefit is to sharpen the regression analysis by considering distinct meteorological regimes separately.

Studying atmospheric data such as ozone concentrations on larger spatial scales raises yet other issues. For the Chicago case studies, it was reasonable to assume that the urban network could be treated as a single coherent region where a single spatial summary, such as the network typical value, made sense. Ozone is well known to have some regional features at spatial scales much larger than isolated urban areas. One challenge in modeling a regional network is to identify subregions within which the variables show reasonably homogeneous behavior. The last case study (Section 4) constructs such subregions from the 118 monitoring station network in Gulf Coast states.

1.2 Data and Dimension Reduction

The statistical tools used to tackle these problems reflect a common theme of data and dimension reduction. This reduction is over time in the case of diurnal modeling, over meteorological covariates in the case of clustering daily meteorology, and over space in the case of identifying subregions. These reductions have clear statistical benefits. For example, studying ozone over a homogeneous subregion may lead to simpler, lower-dimensional models. There is also a more qualitative scientific benefit. The effective reductions often highlight physical processes and mechanisms that might otherwise be missed. For example, from the cluster analysis of meteorology for Houston, one class of high ozone days appears to result from a combination of favorable conditions for ozone formation and the transport of ozone and precursors.

One starting point for reducing data dimension is the lower rank approximation to a matrix based on the singular value decomposition. This basic method helped to identify nonlinear models for diurnal patterns and provided a parsimonious representation for subsequent cluster analysis in the second case study. It is related to the technique of rotated principal components used in Section 4 to identify subregions with similar ozone dynamics.

A more specific tool for dimension reduction comes in the use of additive models for regression analysis. Coupled with nonparametric function estimates, this technique can handle large numbers of variables if interactions among variables are limited. The second case study uses these methods in concert with the meteorological clustering to estimate models for predicting mean ozone levels. This approach is complementary to nonlinear parametric models and to the spatial process models encountered in the preceding chapter.

The reader is referred to the *web companion* for specific data sets and software that are related to the case studies in this chapter.

2 Diurnal Variation in Ozone

The studies described in the previous chapter focused on either the maximum hourly ozone concentration for any given day or the number of days on which a given level was exceeded. The maximum concentration is a reasonable one-dimensional summary of the day's concentrations and is also the focus of regulation. However, for many issues, the daily exposure to ozone would be determined at least in part by other aspects of the day's ozone profile. Figure 1 gives some examples of hourly ozone data for the mean of 11 stations in the Houston metropolitan area and indicates some of the additional structure in daily ozone that is missed by just reporting the maximum value attained. Similarly, Eder *et al.* (1994) have shown that meteorology has a different effect on the *timing* of the daily maximum, for instance, than on its magnitude.

Although ozone monitoring networks provide hourly measurements of ozone, daily summaries of these measurements invariably are used to assess the air quality for a given day. For regulatory applications, the daily maximum 1-hour value was used until 1997, but there are also several other summaries based on running averages or exceedances. These include an 8-hour mean value such as that incorporated in the current NAAQS (see Section 1.2 of Chapter 2), the sum of those values that exceed 60 ppb (SUM06), and the sum of those exceeding 80 ppb (SUM08). While these summaries are useful, it is hard to compare or relate one to another. Also, none of these summaries identifies the time of the maximum value, which plays a role in calculating exposures.

One way to address these problems is to summarize each day's ozone value by fitting a parsimonious daily model to the data for the full day. This is feasible because ozone shows a clear diurnal cycle under conducive meteorological conditions, and to a certain degree, these cycles share some common patterns. This case study, described in more detail in Yang *et al.* (1996), presents a nonlinear model that depends on only three parameters for each day, but can accurately recover the typical summary statistics used in air-quality monitoring mentioned above. Similar to the experience in the Chicago case study, once a nonlinear model has been identified, fitting the model becomes a straightforward exercise. Thus, the emphasis in this section will be to explain how the model was developed using exploratory techniques and using the basic principles of ozone production.

A key initial step in this process was examining the eigenvectors from a *singular value decomposition* (SVD) of a data matrix. This method, described briefly in Section 2.1, is often used to find a reduced-rank approximation; the search for a diurnal model began with this decomposition.[2]

2.1 Singular Value Decomposition

The first tool used for reducing dimensionality in the analyses described in this chapter is the *singular value decomposition*

(SVD); this decomposition of a data matrix is exactly equivalent to *principal components analysis* of the corresponding dispersion matrix.

Routine monitoring of the environment at various hours of the day or at several locations can easily produce volumes of data that are difficult to study or even visualize without first reducing their dimension. It is therefore often helpful to summarize several series of observations into a smaller number of summary series. Sometimes the nature of the data can be used to decide how such summaries should be formed; however, it is also useful to have methods that carry out the process automatically. Singular value decomposition is one of these.

The key steps in the SVD are as follows. First, the observations are written as a matrix – the *data matrix*. The SVD is then used to find a sequence of approximations to the data matrix of ranks $1, 2, \ldots$, where each approximation is the best of its rank, in the least squares sense.

The $(m \times n)$ matrix A has a singular value decomposition (SVD) that can be written as

$$A = UDV^T.$$

Here, U is an orthonormal matrix $(m \times n)$ whose columns are the eigenvectors associated with the appropriate eigenvalues of AA^T; D is a diagonal matrix $(n \times n)$ of non-negative *singular* values ordered from largest to smallest. The square of any singular value is an eigenvalue associated with AA^T or A^TA (both have the same nonzero eigenvalues); V is an $(n \times n)$ orthonormal rotation matrix whose columns are the eigenvectors associated with the appropriate eigenvalues of A^TA. Rewriting the decomposition as a sum of rank-1 matrices gives

$$A = \sum_{j=1}^{m} D_{jj} \mathbf{u_j} \mathbf{v_j}^T.$$

The partial sums provide the reduced-rank approximations referred to above.

One interpretation of the SVD is that linear combinations of the columns of V are used to approximate the rows of A. Often, the eigenvectors associated with the largest singular values give insight into low-dimensional patterns among the the row vectors; this feature is used profitably in identifying a diurnal model. If the singular values decrease rapidly, then an accurate low-dimensional approximation is obtained by retaining only the largest singular

---

[2] The reader should keep in mind that the SVD also played a role in the other two cases studies in this chapter.

values in the sum. A common way to summarize the contribution of each component is the ratio of the square of each singular value to the total sum of squares

$$\frac{D_{kk}^2}{\sum_{j=1}^{m} D_{jj}^2},$$

and can be tied to the fraction of variance explained by the kth component.

The projection of the original data onto the new eigenvector structure of V is given by $S = AV = UD$ and is referred to as the component scores matrix. Omitting columns with small singular values reduces the (column) dimension of the matrix but retains an approximation to the original matrix. For example, in Section 3, examination of the singular values showed that 56 meteorological variables could be reduced to a component scores matrix with 6 columns. This reduction makes clustering algorithms more efficient and reduces storage requirements.

The jth rank-1 matrix in the SVD is defined by the singular value D_{jj} and the vectors $\mathbf{u_j}$ and $\mathbf{v_j}$, each of unit length. We refer to this triple, and equivalently to the matrix $D_{jj}\mathbf{u_j}\mathbf{v_j}^T$, as the jth *mode of variation* in the data (matrix).

Principal components analysis (PCA) simply provides an equivalent means for obtaining the eigenvalues and eigenvectors described above. Note that $AA^T = VD^2V^T$. If A has columns standardized to zero mean and unit variance, the PCA solution obtained from the correlation matrix yields the same eigenvector space as the SVD of this standardized matrix. In PCA, the columns of V are referred to as the *loadings* for the variables.

2.2 Urban Ozone in Houston

The data used in this work are based on hourly ozone records from Houston-area monitoring stations from April to October, 1981 to 1991 ($\approx$ 2300 daily records). There are many stations positioned in the Houston area to monitor ozone concentration. As a prelude to this study, 118 Gulf Coast ozone monitoring stations (located in Texas, Louisiana, Mississippi, Alabama, and Florida) were classified into five spatially cohesive subnetworks using rotated principal components analysis (Section 4.2; see also Royle et al. 1994). The 11 ozone stations in the Houston area which were used in the ozone work reported on below were part of one of these subregions. Modeling of regional ozone will be discussed in Section 4. For this study, the response is the mean hourly ozone for the 11 stations; some randomly sampled 10-day intervals are plotted in Figure 1.

Model Development

In the model development phase, principal components analysis (derived from the singular value decomposition of the data matrix) was used to identify the

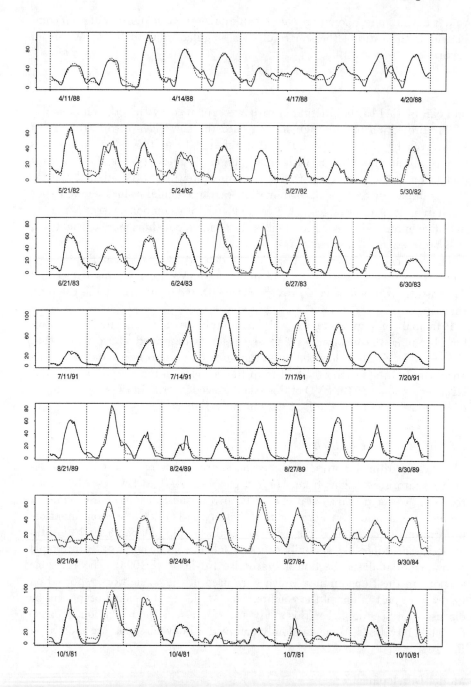

FIGURE 1. Plots of observed diurnal ozone (solid line; ppb) and modeled diurnal ozone (dotted line; ppb) for several, randomly selected, ten day intervals over the years 1981–1991.

basic characteristics of the diurnal profile, represented by the 24-dimensional vector of hourly ozone concentration measurements. The loading vectors for the first four components (i.e. the columns of the V matrix) are shown in Figure 2. The first vector is very similar to the mean profile. The next largest

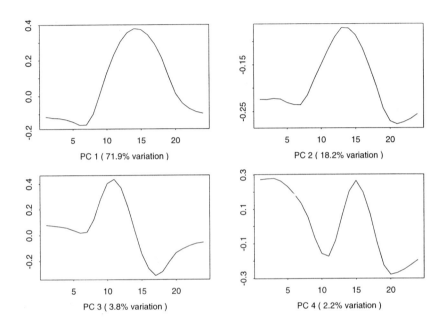

FIGURE 2. Loadings for first four components of diurnal variation.

vector is similar in shape to the first loading. Thus, adding or subtracting this component from a mean profile has the effect of altering the peak height. The next two loading vectors have modes at different locations than the mean curve and so contribute to diurnal patterns by affecting the timing of the peak and changing its shape. Note that the percentage of variation described by each components decreases rapidly after the second loadings.

In summary, the first four loadings explain a substantial amount of the variability in ozone (greater than 95%), suggesting that the principal modes of variation are:

- the overall scale of variation over the day;

- the ratio of the maximum to the minimum; and

- the location of the peak.

These observations suggested the following tentative model:

$$y_{d,h} = \mu_d + \alpha_d M(h + \beta_d),$$

where $y_{d,h}$ is the ozone level at hour h on day d and M is a parent diurnal pattern. When the day d is fixed, the model describes the profile of the given day with three parameters. The general level of ozone is characterized by μ, while α controls the flatness of the profile and β specifies the phase departure from the normal or average profile.

Using a single shift for a whole day appears to be overly rigid. A two hour delay of the peak does not necessarily result in the same amount of delay for the early morning low point. In fact, although there is a noticeable shift for the peak, the timing of the morning minimum appears to be relatively stable. Thus, a shift term $\beta_d \tau(h)$ depending on hour of day as well as day is more realistic than a constant shift β_d, which depends only on day.

Given the parent profile M and the shift function τ, the final model uses three parameters to describe the diurnal behavior of ozone:

$$y_{h,d} = \mu_d + \alpha_d M(h + \beta_d \tau(h)) + \varepsilon_{d,h}.$$

The functions τ and M are called the *shift* function and the *shape* function, respectively. In this model, M has been normalized to 1 at its maximum value and 0 at its minimum value. Similarly, τ is standardized so that it is 1 when M is maximal. In this analysis, M is obtained by rescaling the mean profile over the entire time period. With these normalizations, the model parameters μ_d, α_d, and β_d have clear interpretations. The daily maximum is given by $\mu + \alpha$, where α is an indicator of flatness. When $\beta_d = 1$ (-1) the profile maximum has been shifted one hour earlier (later). An iterative nonlinear least squares procedure was used to find the estimates of μ_d, α_d, and β_d, as well as the shift function τ.

Accuracy of Model Predictions

The fitted daily profiles exhibited diurnal and seasonal variability that one expects in ozone levels. There are, however, days when the profile does not match the observed diurnal ozone pattern, especially when ozone concentrations were low. On those days that were meteorologically favorable for ozone formation, the median value of R^2 was 0.948, while on those days not considered favorable for ozone formation, the R^2 value was under 0.70. For obvious reasons, low ozone days are of less interest than high ozone days. Figure 1 shows model performance for seven ten-day periods.

Figure 3 indicates the close agreement between daily summaries based on the observed (raw) data and those calculated from the model. Because this is a regression model, the daily maximum will be underestimated.[3] Overall, the

---

[3] Estimates of the maximum could be improved through a bias correction based on the fitted regression model in addition to the estimated mean curve.

actual statistics are accurately recovered by the diurnal model, particularly the 8-hour average.

2.3 Conclusions

The results of this section indicate that the method is capable of summarizing each day's ozone measurements into three

statistics, measuring quantities of interest such as the overall intensity of ozone's temporal behavior over the day, the strength of the peak relative to the overnight minimum, and the timing of the peak. The effectiveness of these summary statistics for classifying days and as the starting point for other analyses of the effects of meteorology on ozone levels remains to be studied.

3 Meteorological Clusters and Ozone

For the analysis of ozone concentrations in the Houston area, a primary goal was to build models that were based on physically interpretable weather patterns. The methods used for the Chicago data (Chapter 2) could also have been applied at this location. The GaSP model was refitted and provided results comparable with those from Chicago. The nonlinear parametric model was reestimated for Houston but with unsatisfactory results. Modification of the model was not attempted, but would presumably have led to some improvement. The widely disparate meteorological regimes of the Gulf Coast area favored an approach based on an identification of the unique regimes of the area. The present case study, described in more detail by Davis *et al.* (1998) uses clustering methods to achieve that goal.

This case study draws substantially on the techniques in Eder *et al.* (1994) as applied to the Birmingham urban area. They initially performed principal components analysis (see Section 2.1) on 3-hourly meteorological data. A reduced number of component scores were used in a two-stage cluster analysis procedure. The objective was to determine synoptic-scale meteorological scenarios to which each day during the ozone season could be assigned. Then, the relationship between ozone levels and various surface and upper-air meteorological variables was determined within each cluster using linear regression procedures.

One difference between Eder *et al.* (1994) and the Houston study was the use of more flexible statistical procedures for the within cluster modeling of ozone. There was some concern that the improvement seen by Eder *et al.* in fitting clusters separately as opposed to modeling the full data set could be due to the fact that linear models were used in both cases. Although a linear model applied to the full data may not work well, nonparametric techniques may have the flexibility to adjust for different meteorological regimes; thus their performance could be comparable to the models using clusters.

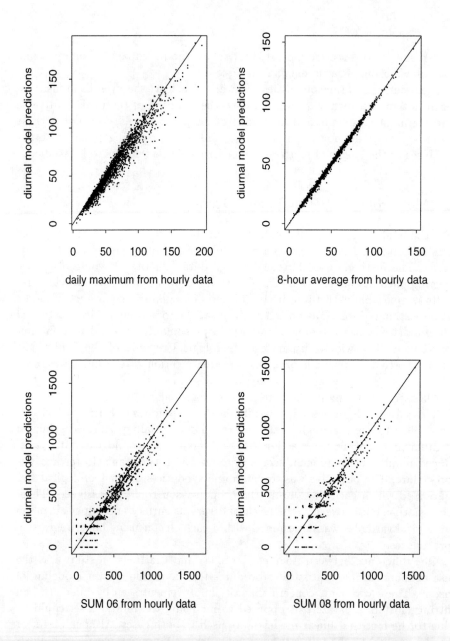

FIGURE 3. Scatterplots of the summary statistics calculated from hourly Houston ozone data and the same statistic based on the diurnal model.

Before describing the results of this case study, it is appropriate to give some background on cluster analysis and nonparametric regression with additive models.

3.1 Cluster Analysis

Clustering can be defined as the grouping of objects that have similar characteristics. Objects that reside in a given cluster are assumed to have a high degree of natural association among themselves, while the individual clusters themselves are assumed to have relatively distinct properties. Although there are a number of clustering procedures, the following two have been found to be quite useful for meteorological data. For a general reference on cluster analysis, see Hartigan (1975).

Hierarchical Cluster Analysis

The starting point for the clustering algorithm is a distance matrix expressing the pairwise distance between all the data vectors. These distances usually use a Euclidean metric. Also, there are several algorithms to aggregate data points into clusters: single linkage, complete linkage, and average linkage. In the *single linkage method*, clusters are formed from individual entities by combining nearest neighbors. In this context, nearest neighbor indicates smallest distance or greatest similarity. *Complete linkage*, while similar to single linkage, has one major difference: At each stage of the computations, the distance between clusters is determined by the distance between two members, one from each cluster, that are most distant from each other. The advantage of this procedure is that all members of a given cluster are within some maximum distance of each other. In the *average linkage procedure*, the distance between two clusters is calculated as the average distance between all pairs of members, where one member of a pair belongs to each cluster. See Johnson and Wichern (1992) for the computational procedures. Kalkstein *et al.* (1987) found that the average linkage method produced the most physically realistic clusters when applied to meteorological data. In any cluster analysis work, one must decide on the number of clusters to retain. This topic is discussed in Eder *et al.* (1994).

k-Means Cluster Analysis

In *k-means* analysis, the objective is to minimize the within-cluster sums of squares. This is achieved by moving observations from one cluster to another, based on the cluster centroid to which the individual member is closest in Euclidean distance. The centroid for the cluster receiving the new member and for the cluster losing the member is then recalculated. Member movement stops when no more reassignments occur, or when some user-defined limit has been reached (Eder *et al.* 1994). Computational procedures are outlined in Johnson and Wichern (1992). One difficulty with applying k-means is the necessity of starting values for the cluster centers. Davis and Kalkstein (1990) suggested using the output from the hierarchical clustering as input to a k-means clustering routine.

3.2 Nonparametric Regression

In many instances, linear or nonlinear regression analysis based on a parametric model is not appropriate: see examples in Härdle (1990), Hastie and Tibshirani (1990), Green and Silverman (1994), and Fan and Gijbels (1996). An alternate procedure to the parametric approach, which assumes that the functional form is fully specified by a finite number of parameters, is to try nonparametric estimation without regard to any specific form. The main advantage to the nonparametric approach is that it provides a flexible method for exploring the relationships among a set of variables.

Splines

A spline function can be represented in the form

$$f(x) = \sum_{j=1}^{q} \beta_j \varphi_j(x), \qquad (3.1)$$

where q is the number of basis functions, the β_j are the model parameters to be estimated, and the $\varphi_j(\cdot)$ are the spline basis functions. The basis functions are usually piecewise polynomial functions with good approximation properties and compact computational forms. It is important to distinguish between regression splines, where the coefficients are found by standard methods, such as least squares and smoothing splines, where the estimate solves a variational problem.

First, we describe basis functions for cubic splines. Assume one has real numbers $x_1, \ldots, x_n$ on an interval $[a, b]$, where $a < x_1 < x_2 < \cdots < x_n < b$. Each φ_j is a piecewise cubic polynomial defined on the interval $[a, b]$. On each subinterval $((a, x_1), (x_1, x_2), \ldots, (x_n, b))$ in $[a, b]$, φ_j is a cubic polynomial, and these polynomial sections fit together at the endpoints (x_i, called "knots") in a manner such that φ_j and its first and second derivatives are continuous at each knot, and thus also on $[a, b]$. An explicit representation for these basis functions with equally spaced knots is given in Chapter 6. A cubic spline is *natural* if its second and third derivatives are zero at a and b.

Given the spline basis, the simplest approach is to estimate the coefficients as conventional parameters. This makes is possible to use standard regression software and simplifies testing for the significance of a variable in an additive model.

As an alternative approach, a cubic *smoothing* spline is the function that minimizes

$$S(g) = \sum_{i=1}^{n}(Y_i - g(x_i))^2 + \alpha \int_a^b (g''(x))^2 dx$$

over all twice-differentiable functions $g(\cdot)$. Here, $\alpha \geq 0$ is a (fixed) smoothing parameter. The basic result for smoothing splines is that the solution has the same form as (3.1) using a natural spline basis with a knot at every observed x value. The use of a roughness penalty term guarantees that the "price" (i.e., $S(g)$) for using a particular curve is determined by its roughness ($\int (g''(x))^2 dx$) in addition to how well it fits the data, as specified by the residual sum of squares, $\sum_{i=1}^{n}(Y_i - g(x_i))^2$. For a given value of α, minimizing $S(g)$ will provide the best compromise between smoothness and goodness-of-fit. Values for α can be estimated from the data using cross-validation. A more general discussion of splines, their computation, and the analysis of data are included in the FUNFITS manual in Appendix A.

The roughness penalty technique can be extended to multiple regression in the form of a semiparametric model where the linearity assumption is suspended on just one of the covariates:

$$Y = g(x) + z^T \beta + \varepsilon.$$

Here $g(\cdot)$ is a smooth curve to be estimated, β is a vector of parameters also to be estimated, and z is a vector of explanatory variables. This approach can be extended to allow for all the covariates to be nonlinear, which is then a generalized additive model (GAM).

Loess

Loess (Chambers and Hastie, 1993) is a smoother based on fitting a linear model to subsets of the data. The fitted value at x_0 is obtained from the weighted least squares fit of y to x. The key to the procedure is that the weights are constructed to be large in a neighborhood of x_0, but decrease to zero outside of this neighborhood. Because this estimate is based on fitting (small) subsets of the data around each point for evaluation, the method can handle large data sets and is a useful tool for exploratory analysis.

Generalized Additive Models (GAM)

The basic additive model is defined by the equation

$$Y_t = \sum_{i=1}^{p} f_i(x_{it}) + \varepsilon_t,$$

where p is the number of covariates and x_i plays the role of a simple linear predictor in ordinary least squares. As in ordinary least squares, $E(\varepsilon_t) = 0$ and $\text{Var}(\varepsilon_t) = \sigma^2$. The $f_i(\cdot)$ terms are arbitrary univariate functions with an $f_i(\cdot)$ modeled for each covariate (Hastie and Tibshirani 1990). Note that in the linear case, $f_i(x_i)$ is equal to $x_i \beta_i$.

An additive model is a compromise between a linear model and a function without any restrictions on the form of the interactions among the independent variables. The motivating principle here is the "curse of dimensionality" (see, e.g., Hastie and Tibshirani 1990, p. 83): The difficulty in estimating a function increases very rapidly as one adds more variables into a model.[4] Although it is simple to estimate a two-dimensional surface, adequately estimating, say, a full 10-dimensional surface to comparable accuracy may require an astronomical sample size. Additive models avoid the curse by restricting the interactions among the variables. Mixed partial derivatives of the functions are assumed to be zero and, in this way, a high-dimensional surface can still be estimated with modest samples sizes.[5]

Almost any smoother such as a spline or *loess* can be used to estimate the individual functions in the model. Besides the choice of smoother, there is also flexibility in specifying the degree of smoothing. In this case study, the smoothing parameter or neighborhood size has been fixed according to the effective number of parameters associated with each function estimate (Hastie and Tibshirani 1990). An alternative method is to choose the smoothing based on the data using cross-validation.[6]

As an alternative to using a smoother to estimate the additive functions, one can apply parametric methods (e.g., orthonormal polynomials, spline basis functions) fitted by least squares rather than the iterative procedures needed to achieve nonparametric fits. As Hastie and Tibshirani (1990) point out, the calculation of standard errors, determination of degrees of freedom, and tests of significance are straightforward for this parametric case, but not so for the nonparametric case, where reliance is placed on less concrete measures.

3.3 Urban Ozone in Houston

Data

The meteorological data used in constructing the clusters consisted of observations made every three hours (01, 04, 07, ..., 22 LDT) recorded at the Houston International Airport. The data cover the period April 1 to October 31 from 1981 to 1992 (2568 days) and include station pressure (mb), temperature (°C), specific humidity (g kg^{-1}), total cloud cover (%), u wind component (m s^{-1}), v wind component (m s^{-1}), wind speed (m s^{-1}). Thus the original data matrix was 2568 (days) × 56 (seven variables each available eight times a day). This matrix was standardized[7] (except for the wind component data) by column before the SVD was applied.

The period of record and suite of meteorological variables used in developing

---

[4]For example, the number of polynomial terms up to degree m in $\Re^d$ is $\binom{m+d}{d}$ and so increases rapidly in the dimension. Compare this to just the $d+1$ terms in a linear function.

[5]The additive model is not just restricted to sums of univariate functions and could include low-dimensional surfaces for variables with strong interactions.

[6]The S-PLUS function addreg in FUNFITS (Appendix A) will determine the smoothing parameters by generalized cross-validation.

[7]Centered and scaled to have zero mean and unit variance.

the statistical models were somewhat different from those used in developing the clusters. Rather than considering all eight of the 3-hourly surface variables observed daily for incorporation into the model, either the daily maximum (for surface temperature) or the daily mean (for the u and v wind components, wind speed, station pressure, specific humidity, and cloud cover) of these observations was used. Solar radiation data ($MJ\,m^{-2}$) integrated over the day were also incorporated into the statistical models. Unfortunately, the radiation data were only available for the 1981–1990 period; as a result, 1991 and 1992 were not used in development of the within-cluster models. Upper-air data from the Lake Charles rawinsonde station, located approximately 200 km to the east-northeast, were also used in the models and consisted of the 850 mb (00 UTC, 12 UTC) temperature, dew-point depression, u and v wind components, and wind speed as well as estimated mixing depth.

The ozone data are based on 11 stations within the Houston, Texas Metropolitan Statistical Area (MSA). The network summary was similar to that used in the Chicago study.

Estimating the Clusters

The first step was to reduce the full data matrix to a more manageable dimension. Based on examining the sizes of the singular values and the meteorological context for this analysis, the first 6 (out of a total of 56) eigenvectors (accounting for 77.6% of the variation) were retained from the SVD of the meteorological covariates. [8] The resulting loading matrix S (2568 × 6) then served as the input matrix to the average linkage clustering routine. [9]

Selection of the final average linkage solution was based on pseudo-F and pseudo-t^2 statistics, both of which indicated seven clusters. The pseudo-F is derived from an analogy with the usual F statistic from analysis of variance and is the ratio of the variance based on *within*-cluster sums of squares to the variance based on *between*-cluster sums of squares. The difficulty in testing for the number of clusters using this statistic is that the reference distribution does not have a single simple form. One problem is that, unlike ordinary ANOVA, membership of a data point in a cluster is based on the observed data values and so the within-cluster variance is smaller than than would be expected using the usual F distribution. Typically, the pseudo-F is plotted as a function of cluster size and the first local minimum of this function is taken as an indicator of a useful number of clusters.

One advantage in the present application is that selecting the number of clusters can also be guided by the climate for Houston and qualitative under-

---

[8] The graphical procedure used to identify the cutoff is known as a Scree test (see Eder *et al.* 1994) and is based on subjectively finding a "knee" in the plot of singular values.

[9] Up to an orthogonal rotation, S is the rank 6 approximation to the full data matrix. Because subsequent clustering is based on Euclidean distance, the same distances are obtained whether the final orthogonal rotation is omitted or included.

standing of the large-scale weather patterns that influence this area. Thus, one would expect the number of distinct meteorological regimes to be on the order of 5 to 10 and so this prior information also served as a guide in choosing the number of clusters. One confirmation of this process is in the next section, where we will see that the clusters are readily interpretable and also imply different mean levels of ozone. This seven-cluster solution was also used as input for the convergent k-means analysis. Upon completion of the two clustering procedures, each day of the 2568-day study period has been grouped with other days exhibiting similar meteorological conditions.

Relationship of the Clusters to Ozone Levels

Because of the critical impact of meteorological conditions on the formation, scavenging, transport, and deposition of ozone, the homogeneous meteorological clusters defined by each clustering technique should exhibit significantly different concentration characteristics. This hypothesis was examined using several statistical tests designed for multiple comparisons of several group means. We should emphasize that the clusters have *not* been constructed using ozone measurements and, in contrast to testing for the number of clusters, these simple statistical tests are valid. For example, the Waller-Duncan K-ratio T test (Steel and Torrie 1980) indicated that the majority of clusters have significantly different mean ozone concentrations. A better segregation of mean concentrations was found with the two-stage clustering approach. Of the $21 = \binom{7}{2}$ possible cluster comparisons, 17 of the k-means pairs have statistically different mean ozone concentrations, while only 13 of the average linkage combinations are statistically significant. Because the k-means results give a sharper delineation among ozone levels these results will be emphasized in the following sections.

Interpreting the Clusters

The characterization of the meteorological clusters and subsequent analysis of ozone concentrations observed within each cluster revealed many intriguing features (Figure 4). Here, we will concentrate on the results for k-means. With both clustering approaches, the three highest ozone concentrations are associated with those clusters identifying anticyclones, which is not unexpected and is supported in numerous other studies. What is unusual about these findings is that the very highest concentrations were associated with the *back side* of the migrating anticyclone,[10] which occurs most often during the months of April and May and again during September and October (this is k-means cluster 3 with mean ozone 88.69 ppb). During these periods, temperatures are considerably cooler than one might expect with high ozone concentrations.

---

[10]This characterization is based on the summary statistics of the meteorological variables within this cluster and examining for days a sample of weather maps assigned to this cluster.

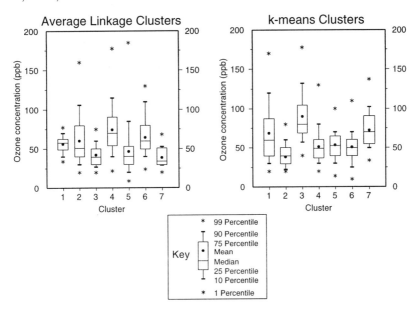

FIGURE 4. Ozone concentrations by cluster.

In this cluster, cloud cover is low, and winds are initially light and from the northeast but eventually turn to the east-southeast. These characteristics suggested that transport of ozone and its precursors may be responsible for the high concentrations.

The next highest concentrations are found with the *front side* of the migrating anticyclone (cluster 7 with mean ozone of 71.74 ppb). This cluster occurs exclusively during the months of April, May, September, and October, and exhibits the coolest temperatures of all clusters. Winds associated with the cluster are from the northeast and cloud cover is low.

The third highest concentrations coincide with the quasi-permanent *Bermuda High* (cluster 1 mean ozone 67.52 ppb), which occurs most often during the summer. Temperatures are the highest of all clusters and cloud cover is rather low, which would contribute to the high ozone concentrations. However, in this cluster, winds are from the southeast, off the Gulf of Mexico, inhibiting transport from the petrochemical corridor and thereby preventing ozone values from becoming exceedingly high.

Within-Cluster Regression Results

The associations hypothesized above were studied in each of the average linkage[11] and k-means clusters. Initially, GAM was used to explore the nature of the association, using both spline and *loess* components. Guided by the results, a model was then constructed using stepwise inclusion of natural spline functions of chosen variables. For the k-means clusters, the R^2 values ranged from 0.48 (cluster 3) to 0.71 (cluster 6), while the residual standard errors ranged from about 9.32 ppb (cluster 6) to 23.39 ppb (cluster 1). In comparison, the GAM fit to the full data had an R^2 of 0.54 and a residual standard deviation of 21.07 ppb.

Three covariates (maximum temperature, v component of the wind, and solar radiation) stand out as being important in nearly all clusters. The results also suggest features that would be difficult to discern from a single monolithic regression model. For example, in the meteorological regimes dominated by anticyclones, all three surface wind covariates (speed, u and v components) were universally significant, indicating the likely importance of transport within these anticyclonic, ozone-conducive clusters. This kind of detail may be useful in improving numerical/physical models (see Chapter 4) for ozone.

One final question naturally arises as to the benefits of modeling within clusters as opposed to considering the data set as a whole. An F-test indicates that there is some advantage in the GAM modeling based on the clusters. However, this advantage is modest, and given the flexibility of GAM, it may be difficult to substantially improve predictions based on clustering. One practical issue is that the residual variance differs significantly among the clusters. Thus, the cluster approach would be preferred in order to account for the apparent heteroscedasticity in the full data.

4 Regional Variation in Ozone

The ozone studies in both Chicago and Houston focused on small-scale networks of monitoring stations. The full NAMS/SLAMS ozone monitoring network is comprised of more than 800 stations, and modeling individual stations or small groups is not practical. In addition, as was noted in Chapter 2, Section 3, ozone also has regional-scale variations that influence both rural and urban areas. These features are missed in considering each monitoring station or a small urban area separately. Thus, one way of reducing the dimension of the monitoring data and to understand larger-scale features in ozone is to consider regional summaries. However, over large areas, neither ozone nor meteorology behaves homogeneously, and the first stage of a study of large-scale variability is to find contiguous spatial patches within which the variations are

---

[11]Because of its limited size, cluster AL 7 was excluded from the modeling analyses.

reasonably homogeneous. This case study illustrates statistical tools to group stations into coherent subregions.

In prior work, Eder et al. (1993) used *rotated principal components analysis* (see Section 4.1) to find such subregions for the eastern United States. In a subsequent NISS study, Royle et al. (1994) used similar methods on data from the Gulf Coast states, one of their subregions forming the Houston network analyzed above in Section 2.

4.1 Rotated Principal Components

Principal components analysis (PCA) was introduced in Section 2.1. It is essentially just the singular value decomposition of the data matrix $A = UDV^T$ and is typically used to find an approximation of reduced rank r, $A_r = U_r D_r V_r^T$. However, this representation of A_r is not unique. If T is any $r \times r$ orthogonal matrix, then A_r also satisfies

$$A_r = U_r D_r T T^{-1} V_r^T = (U_r D_r T)(V_r T)^T.$$

Since this incorporates a new score matrix $U_r D_r T$ and a new loading matrix $V_r T$, it may give different insights into the nature of the approximation. One strategy for choosing T is to introduce as many zero entries as possible into $V_r T$. This is desirable because a zero entry indicates a variable that does not contribute to a particular component. To accomplish this, Kaiser (1958) introduced the Varimax criterion. In its simplest form, this criterion seeks to maximize the sum of the row variances for the squares of the entries of $B = V_r T$.[12] Define σ_j^2 by

$$m^2 \sigma_j^2 = m \sum_{i=1}^{m} \left(B_{i,j}^2\right)^2 - \left(\sum_{i=1}^{m} B_{i,j}^2\right)^2.$$

The rotation matrix T is found by maximizing $\sum_{j=1}^{r} \sigma_j^2$.

When, as here, the rows of B are identified with spatial locations, this means that the components reflect groupings of locations, and if these happen to be contiguous, they may be interpreted as regional or subregional groupings. This makes spatial interpretation easier (Horel 1981). Note that rotated principal components do not explicitly incorporate the relative spatial locations of the ozone stations; thus, an analysis that produces contiguous groups confirms the validity of the approach.

---

[12] For example, if data values are restricted, say, to the interval $[0, M]$, then the sample variance is increased as the values move to the endpoints of this interval. The maximum is achieved with half the values at zero and the other half at M.

4.2 Gulf Coast States

Introduction and Data Considerations

Royle et al. studied ozone concentration observations made at 118 monitoring stations in the Gulf Coast states (Texas, Louisiana, Mississippi, Alabama, and Florida). The first issue in the analysis was missing data. Principal components analysis (PCA) requires a complete matrix and can, therefore, be applied only to a submatrix with no missing observations. To make the analysis as extensive as possible, the number of stations should be large. To make the analysis stable, the number of observations should be large. These goals are in conflict. The authors struck a compromise with a subset of 20 stations having 757 days in common. Following the pattern established by Eder et al. (1994), the authors used a rotated analysis to identify five spatially cohesive subregions. A further 49 stations were judged to have enough data in common with the 20-station network (the criterion was 100 days) to allow each to be assigned to one of these five subregions.

Spatial Analysis

In the Royle et al. analysis, the first six modes of the PCA explained a total of 79.2% of the variation in the 20-station network. The Kaiser rotation was then applied to these six modes. Five spatially cohesive subnetworks interpretable respectively as "Houston," "Dallas," "Louisiana," "Eastern," and "Southwestern" were identified (see Figure 5).The sixth mode weighted heavily only on a single station west of Houston. The ozone concentrations among stations within subnetworks was seen to be more similar than ozone among stations across subnetworks, and the correlation between stations among subregions decreased approximately inversely to distance between subregions. This provided justification for the given subnetwork specification and also suggested that strong regional effects play a role in determining ozone characteristics across the Gulf Coast states region.

Temporal Analysis

The five geographically cohesive subregions that these networks represent were found to have different ozone characteristics as defined by seasonality and trend. The seasonality of ozone was distinctly bimodal in all subregions, with peaks occurring in May and July or August for most subregions. This seasonality appeared to be strongest in the Dallas subregion and weakest in the Louisiana subregion.

The estimated trends (not adjusted for meteorology), shown in Table 1, were negative in all subregions and moderately significant except in the Louisiana subregion. The similarity of the standard errors of the estimated trends, despite the different number of stations within each subnetwork, supports the hypothesis that there are underlying dominant regional effects contributing to

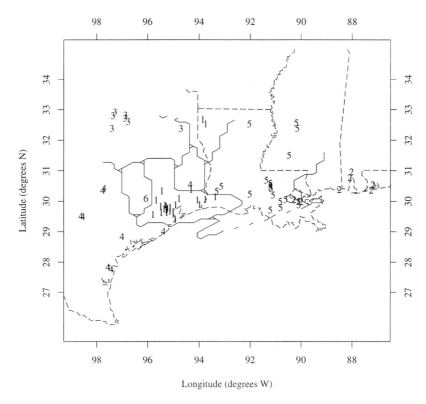

FIGURE 5. Sub region association of 69 ozone monitoring stations.

differences in ozone across subregions.

Conclusions

Decomposition of the Gulf Coast states region into spatially cohesive subregions provides a starting point for the study of ozone in some or all of the individual subregions. Because each of the subregions is more homogeneous

Region	Trend	Std. Error	t Value
Houston	−0.0164	0.00377	−4.36
Eastern	−0.0053	0.00311	−1.72
Dallas	−0.0121	0.00316	−3.81
Southwestern	−0.0069	0.00332	−2.08
Lousiana	−0.0003	0.00323	−0.08

TABLE 1. Subregion trends. Standard errors are adjusted for heteroscedasticity and serial correlation.

with respect to ozone and presumably meteorology than the Gulf Coast states as a whole, construction of meteorologically based modeling exercises will be more interpretable in the context of local meteorological conditions and local ozone transport and production/destruction mechanisms.

5 Summary

In order to determine the effectiveness of the Clean Air Act, scientists have the difficult task of adjusting the trend in ambient ozone concentrations with respect to the influence of meteorology. The group at the National Institute of Statistical Sciences, through collaboration between meteorologists and statisticians, has developed new procedures and extended existing ones in an effort to understand the daily and seasonal ozone cycle. This work has yielded several effective models, based on differing statistical concepts, that can be used to adjust for the effects of meteorology. Results from the classification schemes have provided statistically and physically based rationales for choosing distinctive geographic "influence regimes" for improving the interpretation of ozone air-quality distributions and trends. Likewise, when used in concert with regression analysis, the classification of meteorological regimes better explained the dependence of ozone on meteorology.

6 Future Directions

The new statistical approaches discussed in this chapter and the last have proven useful in the analysis of various aspects of the ozone problem, including trend assessment and forecasting. However, their level of accuracy often falls below that required for many critical decision-making processes, whether they concern forecasting tomorrow's ozone maximum or determining the effectiveness of the Clean Air Act. This shortcoming is often attributable to our lack of understanding and measurement of the fine-scale chemical and physical processes responsible for the formation, transport and scavenging of ozone.

For instance, most statistical models do not account for the complicated diurnal changes that occur in the planetary boundary layer. These and other changes, whether driven by purely local effects, by meso-scale circulation systems such as the sea breeze, or by synoptic-scale features such as frontal systems, are critical and ideally would be incorporated into future models. Newer numerical weather forecasting models, such as the National Weather Service's Eta model with its finer spatial resolution, could also be used. Output from such deterministic models, whether used in an archived sense or in a forecast mode, could provide the requisite meteorological fields needed to achieve a higher level of accuracy for both trend assessment and for predicting ozone concentrations and exceedances.

The challenge then is to construct models, perhaps hybrids of the empirical models discussed in these chapters and the detailed numerical models of meteorology and atmospheric chemistry, that incorporate more of the science of the problem but retain the capability of producing both trend estimates and statements of the uncertainties in their values.

References

Chambers, J.M. and Hastie, T.J. (eds.) (1993). *Statistical Models in S*. Chapman & Hall, New York.

Davis, J.M., Eder, B.K. Nychka, D. and Yang, Q. (1998). Modeling the effects of meteorology on ozone in Houston using clustering and generalized additive models. *Atmospheric Environment* (in press).

Davis, R.E. and Kalkstein, L.S. (1990). Development of an automated spatial synoptic climatological classification. *International Journal of Climatology* **10**, 769–794.

Eder, B.K., Davis, J.M. and Bloomfield, P. (1993). A characterization of the spatiotemporal variability of non-urban ozone concentrations over the Eastern United States. *Atmospheric Environment* **27A**, 2645–2668.

Eder, B.K., Davis, J.M. and Bloomfield, P. (1994). An automated classification scheme designed to better elucidate the dependence of ozone on meteorology. *Journal of Applied Meteorology* **33**, 1182–1199.

Fan, J. and Gijbels, I. (1996). *Local Polynomial Modelling and its Applications*. Chapman & Hall, New York.

Green, P.J. and Silverman, B.W. (1994). *Nonparametric Regression and Gen-*

eralized Linear Models: A Roughness Penalty Approach. Chapman & Hall, New York.

Härdle, W. (1990). *Applied Nonparametric Regression.* Cambridge University Press, Cambridge.

Hartigan, John A. (1975). *Clustering Algorithms.* Wiley, New York.

Hastie, T.J. and Tibshirani, R.J. (1990). *Generalized Additive Models.* Chapman & Hall, New York.

Horel, J.D. (1981). A rotated principal component analysis of the interannual variability of the Northern Hemisphere 500 mb height field. *Monthly Weather Review* **109**, 2080–2092.

Johnson, R.A. and Wichern, D.W. (1992). *Applied Multivariate Statistical Analysis.* Prentice Hall, Englewood Cliffs, NJ.

Kalkstein, L.S., Tan, G. and Skindlov, J.A. (1987). An evaluation of three clustering procedures for use in synoptic climatological classification. *Journal of Climatology and Applied Meteorology* **26**, 717–730.

Kaiser H.F. (1958). The Varimax criterion for analytical rotation in factor analysis. *Psychometrika* **23**, 187–201.

Royle, J.A., Bloomfield, P., Nychka, D. and Yang, Q. (1994). Description of the Gulf states ozone monitoring network and decomposition into subnetworks. Technical Report 22. National Institute of Statistical Sciences, Research Triangle Park, NC.

Steele, R.G.D. and Torrie, J.H. (1980). *Principles and Procedures of Statistics.* Second edition. McGraw-Hill, New York.

Stewart, G.W. (1973). *Introduction to Matrix Computations.* Academic Press, New York.

Yang, Q., Royle, J.A., Nychka, D. and Bloomfield, P. (1996). Diurnal ozone profile modeling. *Proceedings of the American Statistical Association, Section on Statistics and the Environment*, 43–48.

Design of Air-Quality Monitoring Networks

Douglas Nychka
North Carolina State University

Nancy Saltzman
National Institute of Statistical Sciences

1 Introduction

Where should ozone be measured? It is well accepted that high levels of ozone are not only damaging to human health but also reduce crop yield and damage vegetation.[1] However, the continuous measurement of ozone at a location is relatively expensive and so the number and locations of instruments need to be chosen judiciously. This question, although deceptively simple, raises a host of fundamental issues. Most importantly, how can we infer ozone levels at places where measurements are not made? What does it mean to measure ozone well and how many monitoring instruments are really necessary?

1.1 Environmental Issues

Although we will find that these questions, and some answers, can be formulated in terms of statistical and mathematical concepts, this work is rooted in the practical problem of adequately monitoring the environment with limited resources. Figure 1 indicates the locations of the ozone monitoring network for a portion of the Midwest. This chapter is about tackling some specific design problems for this region of the United States. These include *augmenting* a network in size for better spatial representation or *reducing* a network in a way to maintain the best coverage of the region. One intriguing result that emerges from these design case studies is that under certain measures of performance, the monitoring network for ozone is not significantly compromised when the number of stations is reduced by half. The technical contributions from this work are the construction of efficient arrays of stations and the ability to draw objective statistical conclusions about performance. There is currently interest in monitoring other atmospheric pollutants. Reducing the network for a well-studied pollutant, such as ozone, would free up resources to track other constituents.

Given the importance of ozone for air quality, it may seem strange that the design and assessment of regional networks, such as the area in Figure 1, is not already a standard practice. Two reasons for this are the regulatory ozone standards and the difficulty of modeling the ozone field. The location of many

---

[1] See Chapter 1 for a more detailed discussion of the problems with ambient ozone pollution.

stations has been dictated by the regulatory requirements of detecting peaks of very high ozone in urban areas. For example, stations might be placed near known sources of high ozone or high precursor emissions (such as a refinery). To account for the transport of the pollutant, some stations may be sited down-wind of known sources or urban cores. Thus, the care in placement has not been driven by interest in measuring, say, a spatial average, but to have sensitivity for extreme ozone conditions that may occur over a short period of time and at some specific location.

Determining the spatial distribution of ozone is a much more difficult task and is in contrast to the succinct regulatory statistics derived from the network maxima. Ozone has a nonstationary covariance, with the variance and the correlations depending on where it is measured. Typical methods developed in geostatistics do not work well for nonstationary fields such as ozone. Thus, statistical techniques have not been readily available to determine the uncertainty (e.g., standard errors) associated with predictions made from the monitoring locations. But before outlining how these statistical problems might be overcome, it is appropriate to explain how the monitoring data might be used in addition to measuring the regulatory standard.

1.2 Why Find Spatial Predictions for Ozone?

Besides meeting the regulatory requirements, two other applications of the monitoring data will be mentioned: determining exposure of the population and validating numerical (physical) models. First, a spatial average of ozone concentrations could measure the general exposure of individuals to ozone in a region. This measure is in contrast to the regulatory statistics that register extreme events, possibly restricted to a small part of an urban region. Although a spatial average is a reasonable measure, the exact spatial weighting used to form the average is open to debate. For example, a simple (geographic) spatial average of the ozone concentrations over an urban area may ignore the fact that residential areas have lower ozone levels but have more people. This suggests that the spatial weights might be related to population density instead of area and, so, residential areas would be given greater weight. However, places where people work (during the afternoon hours of peak ozone) might be located in core urban areas with high levels of the pollutant. Examples like these suggest that no one particular weighting scheme over space will always be appropriate or accepted. Moreover, the weighting may change over time as activity patterns and the production and transport of ozone in a region change.

Also, an urban area that has poor air quality may be required by law to take steps to decrease the amount of pollution. But how can policymakers decide what measures will be effective? The salient feature of this problem for a statistician is that it *cannot* be answered directly by the analysis of contemporaneous observational data. A natural tool in the decision-making process is to use a physical model that will provide predictions of pollutant concentrations if pollution-generating activities are changed from current practices. In the case of ozone, one needs a physical model describing the creation and transport of ozone as a function of the chemical emissions and relevant weather patterns. Creating a large numerical model for predicting a pollutant, such as the Regional Oxidant Model (ROM) for ozone (Pierce *et al.* 1994; Alapaty *et al.* 1995), is a substantial undertaking and represents a significant scientific contribution. However, in order to use this model for prediction, it is crucial to collect observational data that can validate the numerical model and provide guidance on mechanisms that the modelers have overlooked. Such a comparison is not straightforward and must account for the agreement or disagreement of the model output and observed measurements at different levels of spatial resolution. For example, the model and observed data may be similar if averaged over a 100-square mile block but may have some differences when compared at specific locations. Similar to measures of exposure, we see that model validation also requires a flexibility in the precise way the estimated ozone field is to be used.

1.3 Designs and Data Analysis

The use of monitoring data for measuring exposure or validating numerical models makes it clear that one must be able to predict ozone at spatial locations other than those where measurements are taken. These are complicated problems; the estimated ozone field will be used in many ways, either by being evaluated at particular points or integrated over different regions. Thus, there is no single statistic that will provide all the information needed to solve either problem. A simple way to formalize this flexibility is to demand that ozone be predicted well on the *average* in the design region, or for *all* points in the design region. Those familiar with spatial design (or the design of experiments) will recognize these as criteria based on the prediction variance and suggesting A-optimal (minimizing mean squared error) or G-optimal (minimizing maximum mean squared error) designs (e.g., Johnson *et al.* 1990).

A surprising feature of these studies is that finding the best design according to a specific criterion plays a small role in the project. This is in contrast to a more formal textbook treatment of designs. Why this difference? The pure statistical design problem *assumes* an underlying statistical model for the measurements and so this abstract problem rarely meets data and does not address multiple objectives from the measurements. For the network design problem, the model for ozone must be estimated and so part of this chapter is concerned with the modeling of the ozone field from data. The computation of optimal designs for large numbers of points can be onerous, not only discouraging interactive use but also requiring a large investment in software development. Practical experience (and some theory) suggests that simpler, geometric criteria yield designs that work nearly as well as optimal ones. Also, because the spatial estimates will serve several purposes, applying extensive effort to achieve the optimal design based on a specific criterion may be misguided. The emphasis in the studies discussed below will be on designs that can be readily computed and that fill out the design region. Although the resulting designs will be evaluated according to standard design criteria, we do not expect them to be optimal.

The linchpin for evaluating network designs is to formulate a statistical model for the distribution of ozone over space. We will use some standard spatial statistical methods, but there are some twists required for air-quality measurements such as ozone. In particular, nonstationary spatial covariance functions will be fit to the ozone field. Air-quality data has a time component that can be turned to good benefit in fitting models. This is very different than the usual situation in geostatistics where one has only one "sample" and must estimate the covariance function with only a single observation at each location. Because the covariance model for the ozone field is central to all the design applications, this is presented in a section separate from the case studies. Usually, numerical models such as ROM are validated against observed data. However, for reasons that will be explained below, we will also use the ROM output to enhance the observational record and fit models to it. Turning the tables and using model output for estimating spatial covariance functions is new and not without some controversy.

1.4   Chapter Outline

Because modeling is central to all the design problems discussed herein, the description of the ozone data and the models are abstracted from the case studies of interest and follow in the next two sections. Section 3 ends with a short discussion and an example of design evaluation.

The design problems tackled in this chapter appear in Sections 4, 5, and 6, and are ordered in increasing complexity. They reflect the actual research path on this topic. Also, to give the reader a feeling for the interplay between substantive environmental questions and the development of new statistics, the statistical methodology will be introduced in context as it is needed. The first project is small in scale and investigates *reducing* the size of the 20-station network for the Chicago urban area. By adapting algorithms from regression subset selection, it is possible to search through the large number of possible subsets (2^{20}) to find groups with good properties. The next design problem is to *augment* a region in rural Illinois for better measurement of rural ozone levels. Here, we introduce a tool for constructing designs based on filling space. These designs are not "optimal," but they can be computed rapidly and can adapt to the practical constraints of the geographic regions. Indeed, use of the space-filling designs became the key step in completing this project. The third design problem considers the impact of augmenting or reducing the network for the midwest region depicted in Figure 1. The last section of the chapter draws some general conclusions and suggests areas for future work.

The reader is referred to the *web companion* for specific data sets and software that are related to the case studies in this chapter.

2 Data

In the case studies two forms of ozone data were used in modeling: observational data from the NAMS/SLAMS network, and output from a run of the Regional Oxidant Model (ROM) for the period 6/1/87 – 8/30/87.

2.1 Hourly Ozone and Related Daily Summaries

There are approximately 500 stations in the combined NAMS/SLAMS[2] network that measure ozone in the United States. This network varies in size over time and tends to concentrate stations in urban areas. Figure 1 locates some of these stations for the midwest region of the United States.

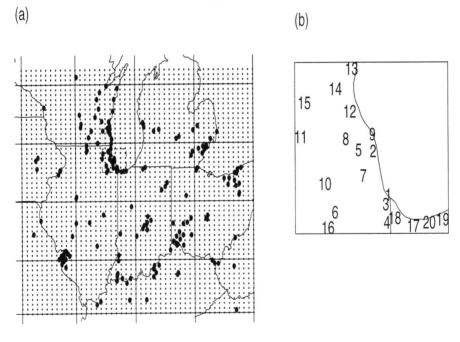

FIGURE 1. (a) locations of the NAMS/SLAMS ozone monitoring stations for 1987 midwestern United States, the ROM grid cell centers, and the 4 × 4 blocks used to investigate nonstationarity of the ozone field (see Figure 8). The second figure (b) is an enlargement of the dashed box from (a), highlighting the the Chicago urban area.

---

[2] The National Air Monitoring System and the State and Local Monitoring Systems. These data are available in the EPA AIRS database. We would especially like to acknowledge the help of William Cox from EPA in making these data available.

To be comparable with the ROM output, the most extensive set of observational data studied were measurements made over the summer of 1987 and consist of approximately 160 stations in the Great Lakes region. In addition, data from the Chicago urban area over the period 1987–1991 were used. Ozone measurements were in units of parts per billion (ppb) and were usually aggregated into hourly measurements.

There are several common ways of converting the 24-hour recording in a day to a daily summary statistic. The main daily summary used in this work is the eight hour average over the period from hours 9 through 17.[3] One advantage of the 8-hour average is that by averaging over time, it can adjust for some of the local transport of ozone. Also, for spatial modeling and prediction, it is helpful for the field to be approximately normally distributed. A statistic based on an average will be more likely to have a distribution that is close to normal. Finally the 8-hour average has relevance given that recent air-quality standards have been phrased with respect to this statistic. Some other useful daily summaries are the maximum and the two truncated sums SUM06 and SUM08. SUM06 (SUM08) is defined as the sum over all measurements that exceed 60 (80) ppb over the day. Note that because of the hard threshold in their definitions, SUM06 and SUM08 can be identically zero for days when ozone measurements are below the threshold.

2.2 Handling Missing Data

Ozone records for a station often contain missing observations and any careful analysis of these data must deal with this problem. For the modeling used in this work, we have filled in missing hourly values using the median polish techniques of Davis *et al.* described in Chapter 2. However, if a substantial number of hourly measurements are missing, the day's record is coded as missing and skipped. In contrast to a time-series analysis where complete time records are very convenient, it is not necessary to fill in the missing day's measurements to derive spatial covariance information. For example, under the assumption that the ozone field is stationary in time, one can estimate the covariance between two locations by the sample covariance based on the available data for these two locations. Although this "pairwise" computation of covariances uses as many data values as possible, it has some problems and more will be said about this in the next section.

2.3 Model Output

It may seem strange to use simulated data (ROM output) to model the covariances of the ozone field. The difficulty with building models for network

---

[3]The more conventional summary is the *maximum* eight hour average found over a 24-hour period. Since ozone levels peak in early afternoon, however, the maximum 8-hour average will usually be over the window from 9–17 hours.

design is that observational data are sparse just in the areas where information is needed. This is the primary reason for considering modifications to the network. To overcome gaps in the observational data, ozone concentrations from the output of ROM are used as surrogate data to model the spatial covariances.

At its most basic form, ROM is a boundary layer photochemical grid model with a resolution that is useful for modeling large (1000 km) domains (Lamb 1983; Young et al. 1989). The model attempts to simulate ozone under different emission inputs and different meteorology, and aids in studying the effect of hypothetical emission patterns on the production of ozone. The model run studied here is from ROM (version 2.2) where weather inputs have been set to follow the "ozone season" for 1987, 6/3/1987 – 8/30/1987 and where emission patterns follow the estimated 1990 inventory. The model output is reported on a grid with a spacing of $1/4°$ latitude by $1/6°$ longitude (18.5 km × 18.5 km) and hourly values for ozone concentrations are available for each grid cell. Model output was validated against observational data (Davis et al. 1998) and the covariance structure was compared over a larger region. The positive results of this comparison were encouraging enough to use the ROM output for modeling.

3 Spatial Models

The key to any method of spatial interpolation or extrapolation is posing a model for the unknown surface. In the context of pollutants, it is useful to assume that the concentrations, or a suitable transformation of them, are a realization of a Gaussian random surface. The reader is referred to Cressie (1991) for background on spatial statistics related to random fields and a review of the literature in this area. The main goal of this section is to review some aspects of spatial statistics and describe some nonstandard details for ozone.

3.1 Random Fields

Formally, let $Z(\boldsymbol{x})$ denote the measured amount of pollutant (or transformed pollutant) at location $\boldsymbol{x} \in \Re^2$. Assume $Z(\boldsymbol{x})$ is normally distributed with $EZ(\boldsymbol{x}) = 0$ and $\text{Cov}(Z(\boldsymbol{x}), Z(\boldsymbol{x}')) = k(\boldsymbol{x}, \boldsymbol{x}')$. Also, let $\boldsymbol{x}_j$ for $1 \leq j \leq N$ be the N locations where the field is observed; $Z_j = Z(\boldsymbol{x}_j)$ and $K_{i,j} = k(\boldsymbol{x}_i, \boldsymbol{x}_j) = \text{Cov}(\boldsymbol{Z}_i, \boldsymbol{Z}_j)$ denote the covariance matrix for this observation vector.

Under the assumption of isotropy,[4] a common covariance function used for a spatial prediction is the exponential function

$$k(\boldsymbol{x}, \boldsymbol{x}') = \sigma^2 e^{-\|\boldsymbol{x}-\boldsymbol{x}'\|/\theta},$$

---

[4]Here, isotropy is taken to mean that the covariance function, k, only depends on $\boldsymbol{x}$ and $\boldsymbol{x}'$ through the distance that they are separated.

where $||\cdot||$ is a measure of distance between two locations. In this chapter, the flat Euclidean distance, $||u|| = \sqrt{u_1^2 + u_2^2}$, is often used for computational convenience, but when possible, great circle distance is used to adjust for curvature of the earth. Note that for any pair of locations separated by a fixed distance, the correlation will be the *same*. In addition, the marginal variances of the fields are assumed to be constant: $\text{Var}(Z(x)) = \sigma^2$. For ozone fields, and many other pollutant fields, this isotropic model is not adequate because the marginal variance, σ^2, is not constant with respect to locations. Indeed, the correlations often depend on the locations as well. Figure 2 indicates the nonstationary nature of the observed ozone correlations. Here, we see that the correlations in the first plot tend to decrease more rapidly with the distance of separation than in the second plot.

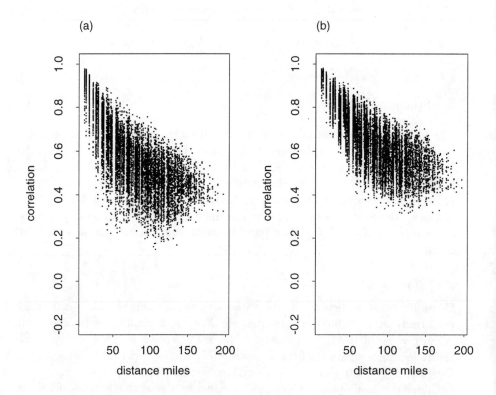

FIGURE 2. Correlograms of ROM output for two blocks for the Great Lakes region. The two plots are correlograms for ROM output in the second and third blocks from the second row from the 4 × 4 arrangement in Figure 1a. Note that these blocks cover the Chicago area, including southern Lake Michigan and the region covering the common borders of Indiana, Michigan, and Ohio, respectively. Pairwise correlations are calculated for pairs of ROM cells based on the 8-hour average over the summer 1987 model run. For each pair of ROM cells in a block, the correlation is plotted against the distance between cells.

A useful extension allows for different marginal variances

$$k(\boldsymbol{x}, \boldsymbol{x}') = \sigma(\boldsymbol{x})\sigma(\boldsymbol{x}')e^{-\|\boldsymbol{x}-\boldsymbol{x}'\|/\theta}$$

but still retains constant (isotropic) correlations. A further generalization is to consider a nonstationary covariance that has a parametric part, such as the exponential model, and a nonparametric covariance, represented as a series of eigenvalues and eigenfunctions:

$$k(\boldsymbol{x}, \boldsymbol{x}') = \sigma(\boldsymbol{x})\sigma(\boldsymbol{x}')\left(\rho e^{-\|\boldsymbol{x}-\boldsymbol{x}'\|/\theta} + \sum_{\nu=1}^{M} \lambda_\nu \psi_\nu(\boldsymbol{x})\psi_\nu(\boldsymbol{x}')\right),$$

where $0 \le \rho \le 1$ and $\{\lambda_\nu > 0\}$. Here, the eigenexpansion is computed from the difference between the sample correlation matrix of the data and the isotropic part of the model. This form is a hybrid between the parametric models typically used in spatial statistics and empirical orthogonal function (EOF) expansions used in the atmospheric sciences (Nychka et al. 1998). This nonstationary model has a simple interpretation in terms of the observed field. Let

$$Z(\boldsymbol{x}) = \sigma(\boldsymbol{x})\left(\rho S(\boldsymbol{x}) + \sum_{\nu=1}^{M} a_\nu \lambda_\nu^{1/2} \psi_\nu(\boldsymbol{x})\right),$$

where $S(\boldsymbol{x})$ is an isotropic process with a marginal variance of one and mean of zero, and $\{a_\nu\}$ are independent $N(0, 1)$ random variables. Thus, the field is a sum of an isotropic process and a linear combination of M additional functions that have random coefficients. The fraction of variance contributed by the parametric part is ρ. A full development of this approach and some background on other nonstationary models can be found in Nychka et al. (1998).

In the standard geostatistics framework where only a single realization of a field is available, these models would be difficult to estimate. This is due to the variability associated with the variogram. Because we are concerned with nonstationary covariances, the individual differences of the variograms cannot be aggregated into distance classes. So, with just a single realization, only one squared difference is available as an empirical estimate of the covariance between two locations. In this work, the data from 69 days are used to estimate the covariance where we assume that the covariance structure is consistent over time. Thus, there are 69 observations each the sample covariance between two points; thus, fitting nonstationary models to the empirical covariances is much more stable.

Based on the data from the summer period in 1987, the two covariance models listed above were fit to the observation data *and* the ROM output. The marginal variances were estimated from the ROM or observational data and interpolated to a grid. For the first model, the range parameter was taken to be the median from fitting exponential models to the 16 blocks of data indicated in Figure 1a. For the second covariance model, the parameter θ and the eigenexpansion were estimated from the middle 69 days of the summer. The remaining 10 days at either end were used for cross-validation in order to determine ρ. The results were $\rho = .5$, $\theta = 140$ (miles) based on the ROM output and $\rho = .25$ and $\theta = 120$ (miles) based on the observational data. For both kinds of data, the number of eigenfunctions in the expansion (M) was fixed (subjectively) at 5 although this parameter might be also estimated by cross-validation. Based on the ROM output field for ozone, the isotropic component is fairly long ranged but only accounts for approximately one-half of the variability in field.

3.2 Spatial Estimates

The spatial prediction problem is quite simple: Determine the pollutant levels at points where they are not observed. More formally, the goal is to estimate $Z(\boldsymbol{x})$ given Z_j for $1 \leq j \leq N$. Under the assumption of normality, the best linear unbiased estimate of $Z(\boldsymbol{x})$ is given by

$$\hat{Z}(\boldsymbol{x}) = \gamma^T K^{-1} \boldsymbol{Z},$$

where $\gamma_j = \text{Cov}(Z(\boldsymbol{x}), Z_j)$ and this estimate has prediction variance

$$E(Z(\boldsymbol{x}) - \hat{Z}(\boldsymbol{x}))^2 = \text{Var}(Z(\boldsymbol{x})) - \gamma^T K^{-1} \gamma. \tag{4.1}$$

Although this optimality should not be taken too seriously, these estimates have been widely successful for fitting and interpolating spatial data. It should be noted that the simple estimate outlined above can be improved by adding in a low-order polynomial term to account for a general trend, and for this more general estimator, the form for the prediction variance is similar. One problem, however, is that the prediction variance formula is sensitive to misspecification of the covariance. Because in practice the covariance is unknown, this may introduce biases in the reported prediction variances.

Prediction variance or its square root, prediction standard error (PSE), is a useful criterion to measure how well a design covers a region. Indeed, if all the modeling assumptions are met, the PSE is proportional to the width of the confidence interval for the spatial predictions. Thus, a good spatial design will tend to make the prediction variance small at all points in a region of interest.

To interpret the design procedures based on subset selection, it is also helpful to define estimates for a linear combination of the field at a discrete set of locations. For example, let $\lambda(Z) = (1/n) \sum_{k=1}^{n} Z(\boldsymbol{u}_k)$ be the average value of the pollutant over a set of points. Let $\gamma_j = \text{Cov}(\lambda(Z), Z(\boldsymbol{x}_j))$ and $\hat{\lambda} = \gamma^T K^{-1} \boldsymbol{z}$, where $\boldsymbol{z}^T = \{Z(\boldsymbol{x}_1), ..., Z(\boldsymbol{x}_n)\}$. Similar to the point predictions, this Kriging estimate is unbiased and has minimum variance among all linear estimators. The corresponding formula for the mean squared error has the same form as above:

$$E(\lambda - \hat{\lambda})^2 = \text{Var}(\lambda) - \gamma^T K^{-1} \gamma. \tag{4.2}$$

3.3 Design Evaluation

Based on the discussion of the uses of the predicted ozone surface, it is reasonable to focus on selecting designs based on the prediction error either for individual points in the design region or for estimating a spatial average. To understand the graphical summaries appearing in the case studies, it is helpful to give an example using estimated prediction variance surfaces for two spatial designs.

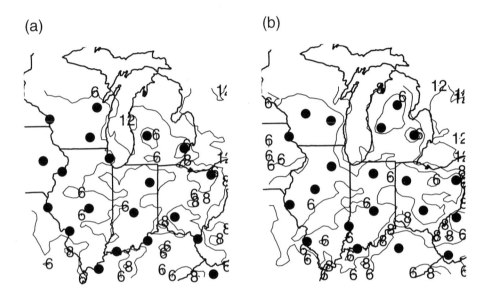

FIGURE 3. A comparison of the prediction errors for two networks consisting of 20 stations. Part (a) is based on selecting a subset of 20 locations from the existing network of 168 (see Figure 1). Part (b) is based on a network of 20 stations selected from a large uniform grid of points. In both cases, the designs were generated by minimizing the coverage criterion $C_{p,q}(\mathcal{D})$, as defined in Section 5.1, with $p = -5$ and $q = 5$.

The first illustration gives contours of prediction standard errors for a 20-point network constrained to be a subset from existing stations, and the second network is generated from a larger candidate set. As might be expected, areas of high prediction variance occur at the edges of the study region, with the constrained network tending to higher PSE.

Besides reporting summary statistics such as the average and maximum prediction variance over the design region, area/variance curves can be used to summarize the fraction of area in the design region with a prediction variance less than a particular value. Examples of these graphical summaries are given in Figure 4 and the difference between the two 20-point designs is shown in

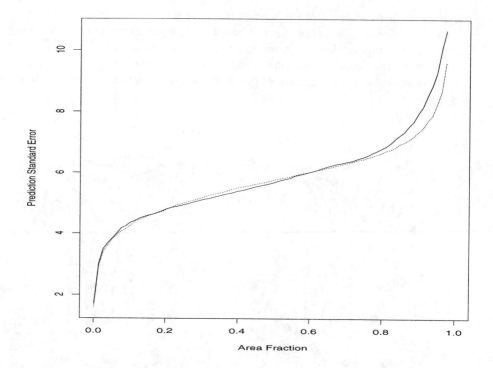

FIGURE 4. The prediction error as a function of the fraction of area of the design region. The solid line is performance of the 20-point constrained network from Figure 3a and the dashed line is performance of the 20-point network from Figure 3b.

this figure. In terms of area, the two networks exhibit similar performance over approximately 80% of the design region. However, for the remaining fraction of area, the PSEs diverge, with the maximum PSEs being substantially different: 10.7 ppb for the constrained network and 9.7 ppb for the unconstrained one.

Another evaluation criterion is based on the concept of thinning a network. Any subset of the network can be used to predict the average for the full (unthinned) network. The variance of this estimate is one measure of how well a particular subset can represent the full network and, conversely, of the efficiency of the original network.

4 Thinning a Small Urban Network

The first design problem studies how the small network for the Chicago urban area might be thinned in an efficient manner. In this study, we will concentrate on the accuracy of estimates of a spatial average; this has a direct connection with exposure measures or other spatial summaries. The basic outline of this work came about by an interest in doing something simple, avoiding the involved modeling of the spatial covariances, but exploiting the time replicates of the data. Also, we wanted to consider several different daily summaries of hourly ozone and their impact on the design choice.

4.1 Preliminary Results

The basic idea for thinning this small set of station information came from an admittedly *ad hoc* exercise using regression on the observed data. For each day, form the average of the daily summaries from the 20 stations and take the 20 individual station values as the "explanatory" variables that can be used to predict the full average. Now, consider the problem of finding a subset of J stations that gives the best linear prediction of the network average. Phrased in this way, the problem is just a regression model

$$Y = \alpha + X_J \beta_J + e,$$

where Y is the network average for each day. X_J is a matrix with the columns being the observed ozone values for a subset of J stations and the rows being the daily summaries for the stations on a given day. An independent variable in this regression equation is identified with a particular station; therefore, selecting a subset of variables is equivalent to identifying a subset of the monitoring network. Initially, we used the *leaps* subset procedure in S-PLUS to find the subsets of different sizes that minimized the residual sums of squares. Some of the resulting subsets of stations are plotted in the first column in Figure 6. Despite the apparent abuse of regression methodology, the results seemed promising. Of course, the fundamental question is: What kind of designs are produced and whether they are robust to several measures of design performance? In order to go any further, it was important to figure out what this procedure was actually doing in selecting a design.

4.2 Designs from Subset Selection

The *ad hoc* selection strategy described above can be justified by making the connection between regression subset selection and minimizing the prediction variance of the estimated average. The basic algorithms for doing this are well known (e.g., Cressie 1991, Sec. 5.9), but the convenience of using off-the-shelf regression software has not been emphasized.

Let λ be the average over the candidate points (or full network) and let $\hat{\lambda}$ be the estimate of this average based on a subset. A good design will estimate λ with a small variance. An important connection is that the mean squared error for $\hat{\lambda}$ can be expressed formally as the residual sum of squares (RSS) from a linear regression. Moreover, regression variables are identified with particular locations. Recall the formula

$$(Y - \hat{Y})^T(Y - \hat{Y}) = Y^T(I - X(X^TX)^{-1}X^T)Y = Y^TY - Y^TX(X^TX)^{-1}X^TY.$$

Thus, if the cross products satisfy the relationships $X^T X = K$, $X^T Y = \gamma$, and $Y^T Y = \text{Var}(\lambda)$, the residual sum of squares from the regression of Y on X will be equal to the mean squared error of the Kriging estimate for λ, Equation (4.2). We can now interpret the regression exercise introduced at the beginning of this section. If the data are centered about their means, then $(1/n)K$ and $(1/n)\gamma$ are just the sample covariances based on the observed data. The subset selection finds the best subset of stations based on these choices for the covariances.

A general form for the regression subset selection problem is to find the solution to

$$\min_{\beta \in \mathcal{M}} (Y - X\beta)^T (Y - X\beta)$$

for some set $\mathcal{M}$. For usual subset selection, if J is the number of nonzero parameter values, then $\beta \in \mathcal{M}$ if $N - J$ values of the parameter vector are zero. We will refer to this estimate of the subset as the *leaps* procedure since it is computed by a leaps and bounds algorithm and is implemented in S-PLUS by the *leaps* function (Furnival and Wilson 1974; Becker et al. 1988). If X and Y are constructed as prescribed above, then minimizing the RSS is the same as minimizing the variance of $\hat{\lambda}$. Moreover, the best regression "subset" corresponds to the subset of candidate points for the spatial design. Given this correspondence, the best regression subset is then the optimal subset of locations with respect to predicting λ.

An alternative strategy for subset selection uses a different constraint set. The *lasso* procedure (Tibshirani 1996) requires that the sum of the absolute values of the parameter vectors be less than a fixed value, i.e.,

$$\mathcal{M} = \left\{ \beta : \sum_{j=1}^{N} |\beta_j| < t \right\}.$$

At first sight, it is not clear why the solution to this constrained problem is related to subset selection. Due to the properties of the absolute value function, however, the constrained solution will force some of the components of β to be identically zero. The number of zeros in the solution can be decreased by increasing t and one identifies the best subset with those variables that have nonzero parameter values. Of course, if t is made large enough, the full least squares solution will satisfy the constraint and thus will also be the *lasso* solution. It is helpful to scale the X variables so that the least squares solution has a constraint of one. Thus, t has a range of $[0, 1]$. Figure 5 illustrates the operation of the *lasso* when it is applied to pick subset designs for the Chicago urban network.

We have found that the *lasso* tends to favor points in the center of the design region and along the edges. The designs based on four- and five- point cluster stations in the center. For this particular example, the sequence of *lasso* subsets are also nested. This is a useful property if one wanted to identify a

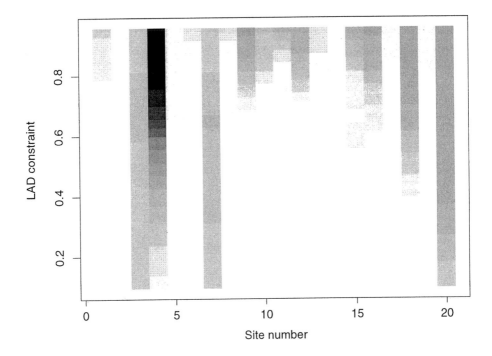

FIGURE 5. Estimated parameters for predicting the full network average using the *lasso* subset selection method. The least absolute deviations (LAD) constraint is the value of the *lasso* equation parameter t as a fraction of its value obtained with the ordinary least square (OLS) parameters. The absolute values of the parameter sizes β_j are represented as a gray-scale value and the site numbers along the horizontal axis can be matched with those in Figure 1b. Stations are eliminated from the solution by decreasing the LAD constraint from one, the OLS solution for all stations.

design consisting of a core of several stations that could then be expanded to include more locations.

Both the *leaps* and *lasso* procedures run fairly quickly on UNIX workstation and so the subset regressions can be computed rapidly. The *leaps* procedure is implemented as part of the interactive spatial design package DI (Nychka *et al.* 1996a); see Appendix B.

4.3 Results

For the period 1987–1991 there were 20 stations in operation in the Chicago urban area (see Figure 1b). The observations recorded for the summer of 1987 were used to estimate a sample covariance for ozone at the network locations and generate designs that were subsets of the full 20-station network. The remaining three years were used to validate the designs and calculate the prediction error. Note that without further modeling, the covariance is only available at the network locations and so the designs were compared based on how well they could estimate the full network average. Figure 6 reports some of the designs found by the *leaps* procedure and the *lasso* for the eight hour average daily summary. The third column of designs in this figure are based on a geometric criterion that is described in the next case study. The designs generated from the summer 1987 data were then validated using the subsequent three years; the results for the eight hour average are summarized in Figure 7. The most important feature is the rapid increase in the design efficiency as the number of sites increases. For example, a design containing five locations appears to do well in estimating the average of the full network.

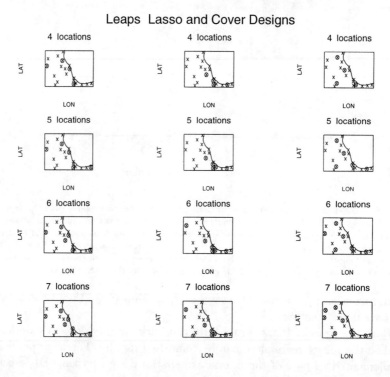

FIGURE 6. Subset designs for the Chicago urban area. The first and second columns are *leaps* and *lasso* designs, respectively, based on the daily 8-hour average over the period 6/3/87 to 8/30/87. In the third column are coverage designs.

Additional stations produce some decrease in the variance, but this improvement is marginal relative to the large decrease up to five stations. In quantitative terms, a five-station network has a prediction root mean squared error of 2.5 ppb. As a benchmark, the reader should note that the unconditional standard deviation for the full network average over the three year validation period is much higher: 16.1 ppb. Using the other daily summaries

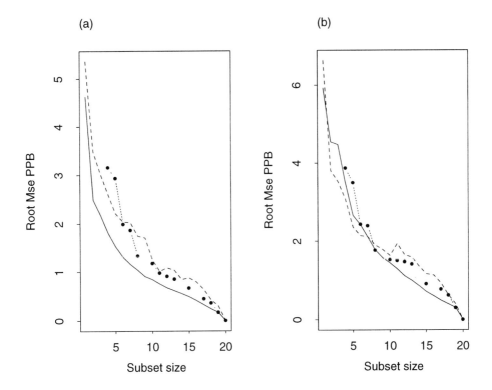

FIGURE 7. Root mean squared error for predicting the full Chicago area network average based on several different subset figures/designs. Plot (a) is the root mean squared error between the predicted average and the full network average observed for each day from 6/3/87 to 8/30/87. The data over this period have also been used to select the figures/designs. Plot (b) is the same comparison, but for the three subsequent summers, 1988, 1989, and 1990: *leaps* designs (solid), *lasso* (points), and coverage (dashed line).

produced different subsets, but similar patterns emerge when the designs were validated.

5 Adding Rural Stations to Northern Illinois

One issue often raised in discussions with EPA scientists was the adequacy of the monitoring network for measuring ozone in rural areas. Few stations are sited in rural areas and a natural question is how much improvement might be expected by adding more locations. A section of northern Illinois was used for testing the methodology of estimating nonstationary covariance functions and generating designs (Nychka *et al.* 1996b). This region coincides with an 18×6 rectangle of ROM grid cells and is part of the validation region in Davis *et al.* (1998) and was chosen because there was already some familiarity with the ozone field in this area. For this work, only the 8-hour average daily summary was used. Although the SUM06 and SUM08 statistics are also relevant, an average was chosen for convenience because of its distributional properties.

The designs for this problem and the larger region in the next section were created using a geometric (space-filling) criterion. This is very different than the usual spatial designs constructed using the covariance function. This choice was made largely on computational grounds. Recall that the prediction variance from Section 3 depends on the inverse of the covariance matrix of the observed points and, so, the objective function to find an optimal design will also include this term. In this case, the number of observations is the total number of stations (163) plus the extra stations (5–10) allocated to the rural study region. Given that the matrix inversion must be done for every evaluation of the objective function, one would expect the minimization of the average prediction error to be computationally intensive. This task is compounded by the fact that typically there are many local minima in the design criterion and, therefore, a conservative strategy is to repeat the optimization with a suite of starting designs.

Note that even though *construction* of the designs are not based on a covariance model, the *evaluation* of the designs are based on their performance for spatial prediction. So if a space-filling design exhibits acceptable properties for, say, average prediction variance, the corresponding A-optimal design must do as well or better. In this way, the space-filling designs in this study set upper bounds on what is possible.

5.1 Space-Filling Designs

A family of design criteria, independent of the assumed covariance function, may be based on geometric measures of how well a given design covers the design region. We will see that these designs are relatively easy to generate and can build in the natural geographic constraints common with environmental problems.

The basic simplification is to reduce the design region to a large, but finite, set of *candidate* points, $\mathcal{C}$. Let $\mathcal{D} \subset \mathcal{C}$ denote the set of N design points. A metric for the distance between any point $\boldsymbol{x}$ and a particular design is

$$d_p(\boldsymbol{x}, \mathcal{D}) = \left(\sum_{\boldsymbol{u} \in \mathcal{D}} ||\boldsymbol{x} - \boldsymbol{u}||^p \right)^{(1/p)}.$$

This metric measures how well the design *covers* the point $\boldsymbol{x}$. For $p < 0$, it is easy to show that $d_p(\boldsymbol{x}, \mathcal{D}) \to 0$ as $\boldsymbol{x}$ converges to a member of $\mathcal{D}$. This relation makes sense because one would expect the design points to cover themselves perfectly. For $q > 0$, an overall coverage criterion is an L_q average of coverage points in the design region

$$C_{p,q}(\mathcal{D}) = \left(\sum_{\boldsymbol{u} \in \mathcal{C}} d_p(\boldsymbol{x}, \mathcal{D})^q \right)^{(1/q)}.$$

The coverage design for a given size is the subset that minimizes $C_{p,q}(\mathcal{D})$ for all $\mathcal{D} \subset \mathcal{C}$. In the limit as $p \to -\infty$ and $q \to \infty$, $C_{p,q}$ converges to a criterion used to define *minimax* space-filling designs. Explicitly, the minimax design minimizes the maximum of nearest-neighbor distances among points in the candidate set to those in the design. It helps to interpret this criterion by analogy to the problem of locating convenience stores in a city. The stores should be sited to be close to the customers and the *minimax* design solution locates stores so that the maximum distance that any customer has to travel to his closest store is minimized. Johnson *et al.* (1990) give some theoretical connections between space-filling designs and those based on prediction error for a spatial process. They show that as the correlations tend toward independence, the minimax designs and those based on minimizing the maximum prediction error are identical.

Coverage designs are generated using a simple "swapping" algorithm. One successively swaps each design point for a candidate point and determines whether the criterion is reduced. If so, the design point is replaced. This is continued until one cannot make any productive swaps. By keeping careful track of how the coverage criterion changes when two points are swapped, one can greatly reduce the number of computations. This is possible because many of the pairwise distances between design points and candidates do not change when a single swap is made. The resulting algorithm is simple and although not gradient based, it has the advantage of being readily implemented in a higher-level language such as S-PLUS. This makes it possible to consider fairly general forms of coverage metrics, because the metric need only be written as a high level function in S-PLUS. Also, no structure is assumed for the candidate set, so complicated and practical constraints can be built into the design. For example, in the Great Lakes region, it is important to keep the monitoring sites on land! This constraint is difficult to parameterize, however, due to irregular lake shorelines. The swapping strategy allows one to simply exclude any point over water from the candidate set. It should be noted that the swapping algorithm will always converge but is not guaranteed to produce a global optimum or even give the same answer from different initial designs. Thus, some care should be used in computing designs using this technique.

An important feature of the distance metric, $d_p(\cdot, \cdot)$ is that at least in a qualitative sense, it is similar to a prediction variance over the design region. Figure 8c is a contour plot of $d_p(\boldsymbol{x}, \mathcal{D})$ with $p = -1$ for a 20-point subset of the NAMS/SLAMS stations. This surface is zero at the design points and has local maxima at points that are in gaps of the network or at the boundaries. The general pattern follows the prediction variance surface based on an isotropic model (Figure 8a). Keeping in mind that the coverage criterion involves some form of averaging over the coverage surface, it is reasonable to expect that averages of the prediction variance should correspond to coverage. Specifically, with $q = 1$, one recovers an average value of the surface, and, of course, when

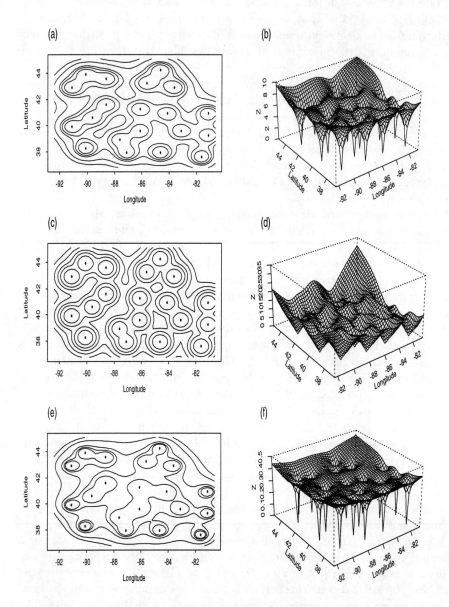

FIGURE 8. A comparison of prediction error and the space-filling coverage function. Parts (a) and (b) are the prediction error surfaces for the 20-point design illustrated in Figure 1a. Parts (c) and (d) are the "distances" between each point in the design region and the design points based on the coverage metric d_p with $p = -5$. The last pair of plots illustrates the surface using a covariance-filling criterion. Here, an exponential function is used to define a metric and the range parameter is the same as that used for the prediction errors in (a) and (b).

$q = \infty$, the coverage criterion is the maximum value of the surface.

One may also modify the coverage metric to be closer to a prediction variance and to reflect the correlation scales associated with a particular spatial process. Quite simply, we take $p = -1$ and replace the Euclidean norm in d_p with

$$k(\boldsymbol{x}, \boldsymbol{x}) - k(\boldsymbol{x}, \boldsymbol{u}) = \sigma(\boldsymbol{x})^2(1 - C(\boldsymbol{x}, \boldsymbol{u})),$$

where $C(\cdot, \cdot)$ is the correlation of the field at two locations. Specifically in the case of the isotropic exponential covariance, a *covariance-filling* criterion with $q = 1$ is

$$\left(\frac{1}{\sigma^2}\right) \sum_{u \in \mathcal{C}} \left[\sum_{x \in \mathcal{D}} \frac{1}{1 - e^{-\|\boldsymbol{u}-\boldsymbol{x}\|/\theta}} \right]^{-1}$$

Figures 8e and 8f are contour and perspective plots, respective of the surface using this modification with a range parameter that is appropriate for ozone ($\theta = 161.3$). At least qualitatively, we see that this metric gives better agreement with the actual prediction variance surface.

5.2 Results for Rural Illinois

Without incorporating stations that border the region, designs will favor the edges of the region to the detriment of adding interior points. To adjust for this effect, the existing monitoring stations were added to the design as fixed points, and new points were selected from within the design region. The candidate set was taken to be an arbitrary grid of 432 (36 × 12) points within the test region. Figure 9 summarizes some of the results of filling in the rural area with additional monitoring sites. In this example, 5 and 10 points were added using the coverage criterion ($p = -5$ and $q = 5$). Because of the interest in filling in this rectangular subregion, the prediction error was only evaluated within this area. Overall, the addition of five points decreased the median prediction standard deviation by 25%, from 3.9 ppb to 3.0 ppb. The addition of 10 points provided a greater decrease, to 2.8 ppb.

6 Modifying Regional Networks

The last design study quantified the results of thinning, augmenting, and creating a new ozone network for the Great Lakes region in Figure 1a. The practical considerations here involved use of the current locations or additions to increase flexibility by changing the sites. The differences between these two choices is illustrated in Figure 2 for two 20-point networks. Here, the disadvantage of using the current monitors is that voids such as the one in northern Michigan, where there are no stations, cannot be filled in. In this study, stations outside this region were not included in the spatial prediction or used

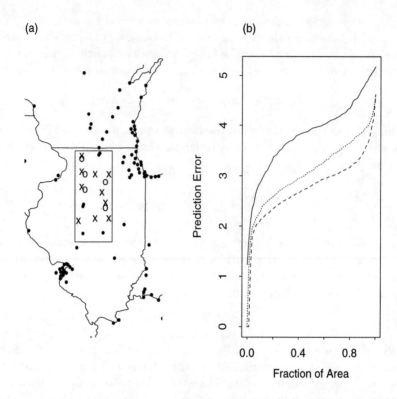

FIGURE 9. Coverage designs adding 5 (o) or 10 (x) additional locations within a rectangular area in northern Illinois. Part (b) summarizes the prediction standard errors for the original network of 168 stations (solid), adding 5 stations (short dashes), or adding 10 stations (long dashes).

as fixed points in the design. Thus the resulting designs could be sensitive to boundary effects. The hope was that designs with a larger number of points might reduce the edge effects caused by ignoring monitoring sites outside this region.

In evaluating the design, it was important to investigate the sensitivity of the results to the covariance function. This requirement is one reason why two different covariance models were fit to observational data and the ROM output.

6.1 Results for the Larger Midwest Network

The current network was increased in size (augmented) in the following way. The existing stations were set as fixed members of the design and the coverage criterion was minimized by adding a specific number of additional locations drawn from a candidate set. As a result, new network locations were placed at the edges of the region rather than filling in voids in the interior. Due to this placement, augmentation had little effect on the network performance except

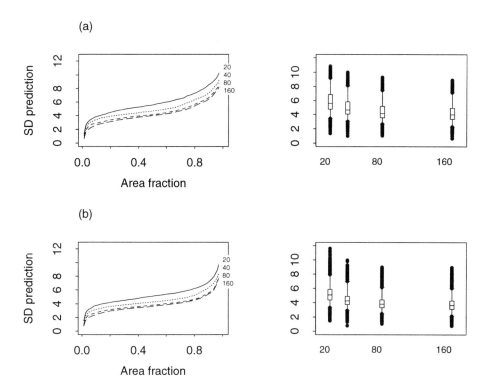

FIGURE 10. A comparison of coverage design ($p = -5$ and $q = 5$) performances for the Great Lakes region based on different network sizes and covariance models. Plot (a) evaluates the prediction error based on a nonstationary covariance function from the observed station data over the summer of 1987. The network sizes are 20, 40, 80, and 160 stations, and the curves are monotonically ordered in network size. Plot (b) is the same summary, except that the covariance function has been estimated from the ROM output. In the second column are boxplots of the distribution of the standard deviations.

along boundaries. One important exception was the placement of stations in northern Michigan, where the results suggested the need for additional sites.

Smaller or thinned networks were constructed by either considering subsets of the existing stations (constrained) or drawing from a larger uniform grid as the candidate set (unconstrained). For small numbers of stations, the performance of the constrained and unconstrained networks was similar. The differences in candidate sets became more apparent as the number of design points approached the total number of existing stations.

The variance/area plots (Figure 10) for different sized networks drawn from the existing NAMS/SLAMS sites were typical of the efficiency of a smaller monitoring network. For example, a subnetwork of 80 stations was comparable to the full network of 168 stations. The size of the prediction variances changed between the covariance models estimated from observational as opposed to

ROM output. This is not surprising given the different values for the range in the stationary component. However, the qualitative features relating the prediction variance to network size appeared to be fairly insensitive to different models for the covariance.

7 Scientific Contributions and Discussion

This research on spatial designs has shown that it is feasible to reduce or augment monitoring networks in an objective manner. Designs that are based solely on geometric criteria often do an excellent job of locating stations. In the absence of covariance information, this promises to be a useful design tool.

By validating against independent data in the case of the Chicago urban network, or by using an estimated covariance based on ROM output, it is possible to assess the prediction performance of the designs. One important physical feature of ozone is the fairly high correlations among concentrations at different locations. This can be quantified by the large range parameter (θ) estimated for the exponential portion of the covariance function. This property results in designed subnetworks having similar predictive performance to the full network. In particular, reducing the number of stations in the Great Lakes region to half the number of stations only inflates the median prediction standard error by about 10%. Practically speaking, a reduction in the number of monitors would have limited effect on the accuracy of the spatial extrapolations since the "error bars" would only be increased by this amount when half the network is used for prediction.

7.1 Future Directions

Trend detection is an important component in pollutant monitoring (Helsel and Hirsch 1988; Esterby et al. 1992; Styer 1994), and has two connections to network design. First, a network that yields more accurate and less biased estimates of regional summaries will reduce the standard error of a trend estimate. Second, a more extensive trend analysis can involve determining the differences in trends of the pollutant level at different locations. In this latter case, one could consider estimating a trend surface for an entire region. This spatial problem is different in character from just spatial prediction of the pollutant. Optimal solutions may require a different network because the spatial covariances for pollutant levels may be different from the the covariances for spatial trends. An example of thinning a wet deposition monitoring network with regard to trend estimates has been studied by Oehlert (1996).

This work clearly demonstrates the value of numerical models for understanding the properties of observational networks. In this role, it is important to validate these numerical models, not only in terms of mean levels but also in terms of covariance structure. This second-order validation for ROM is clearly needed to justify use of an estimated covariance function base on ROM output.

Also, the proposed designs attempt to spread points uniformly over the design region. Although this may be the best strategy for making the prediction variances small, the resulting designs will not contain much information about short- or medium-range covariance structure. Thus, the network may be efficient for prediction but have little power to check a covariance model. In particular, it may not be a good design with respect to validating second-order properties (such as covariances) of a numerical model. An important area of future work is to investigate designs that can also be used to estimate spatial covariance structure at different scales.

References

Alapaty, K., Olerud, D.T. and Hanna, A.F. (1995). Sensitivity of regional oxidant model predictions to prognostic and diagnostic meteorological fields. *Journal of Applied Meteorology* **34**, 1787–1801.

Becker, R.A., Chambers, J.M. and Wilks, A.R. (1988). *The New S Language: A Programming Environment for Data Analysis and Graphics.* Wadsworth & Brooks/Cole, Pacific Grove, CA.

Cressie, N.A.C. (1991). *Statistics for Spatial Data.* Wiley, New York.

Davis, J., Bailey, B., Nychka, D. and Vorburger, L. (1998). A comparison of the regional oxidant model with observational ozone measurements. Technical Report. National Institute of Statistical Sciences, Research Triangle Park, NC.

Esterby, S.R., El-Shaarawi, A.H. and Block, H.O. (1992). Detection of water quality changes along a river system. *Environmental Monitoring and Assessment* **23**, 219–242.

Furnival, G.M. and Wilson, R.W., Jr. (1974). Regression by leaps and bounds. *Technometrics* **16**, 499–511.

Helsel, D.R. and Hirsch, R.M. (1988). Discussion of "Applicability of the t-test for detecting trends in water quality variables." *Water Resources Bulletin* **24**, 201–204.

Johnson, M.E., Moore, L.M., and Ylvisaker, D. (1990). Minimax and maximin distance designs. *Journal of Statistical Planning and Inference* **26**, 131–148.

Lamb, R.G. (1983). A Regional Scale (100 km) Model of photochemical air pollution: Part 1 Theoretical foundation. Technical Report EPA-6000/3-83-035, Environmental Protection Agency, Washington, DC.

Nychka, D., Jonkman, J. and Saltzman, N. (1998). A nonstationary covariance model with application to spatial prediction and network designs. Technical Report. National Institute of Statistical Sciences, Research Triangle Park, NC.

Nychka, D., Saltzman, N. and Royle, J.A. (1996a). Design Interface: A graphical tool for analyzing and constructing spatial designs. Technical Report 42. National Institute of Statistical Sciences, Research Triangle Park, NC.

Nychka, D., Yang, Q. and Royle, J.A. (1996b). Constructing spatial designs using regression subset selection. *Statistics for the Environment 3: Sampling and the Environment*, V. Barnett and K.F. Turkman (eds.). Wiley, New York, 131–154.

Oehlert, G.W. (1996). Shrinking a wet deposition network. *Atmospheric Environment* **30**, 1347–1357.

Pierce, T.E., Milford, J.B. and Gao, D. (1994). Ozone precursor levels and responses to emissions reductions: analysis of regional oxidant model results. *Atmospheric Environment* **28**, 2093–2104.

Styer, P.E. (1994). An illustration of the use of generalized linear models to measure long-term trends in the wet deposition of sulfate. Technical Report 18. National Institute of Statistical Sciences, Research Triangle Park, NC.

Tibshirani, R. (1996). Regression selection and shrinkage via the *lasso*. *Journal of the Royal Statistical Society, series B* **58**, 267–288.

Young, J., Aissa, M., Boehm, T., Coats, C., Eichinger, J., Grimes, D., Hallyburton, S., Olerund, D., Roselle, S., Van Meter, A., Wayland, R. and Pierce, T. (1989). Development of the Regional Oxidant Model Version 2.1. Technical Report EPA-600/3/89-89-44, Environmental Protection Agency, Washington, DC.

Estimating Trends in the Atmospheric Deposition of Pollutants

David Holland
U.S. Environmental Protection Agency

1 Introduction

Atmospheric deposition is widely recognized as a major environmental problem. The release in this century of anthropogenic emissions has polluted the atmosphere and changed its chemistry. Atmospheric deposition of acids and bases occurs through both wet and dry mechanisms. Wet deposition is the result of precipitation removing gases and large and fine particles from the atmosphere. Dry deposition is the transfer of gases and particles to the ground surface in the absence of precipitation. Changes over time in these components of deposition reflect changes in the meteorology and chemistry of the atmosphere, including the effects of emission changes.

In the United States, a focal point of the Clean Air Act from its first passage in 1970 to the amendments of 1990 (CAAA) has been the effects of deposition on terrestrial and aquatic ecosystems. In response to evidence that anthropogenic emissions of sulfur and nitrogen oxides were resulting in increased precipitation acidity and subsequent damage to sensitive ecosystems, the CAAA created the Acid Rain Program to limit emissions of these acid precipitation precursors. One primary goal of this program is a 40% reduction of 1980 sulfur dioxide emissions by the year 2000. To determine whether these controls are having the intended effect, pertinent scientific information must be analyzed to reveal if, when, where, and to what extent the nation's air quality is improving. Toward this end, this chapter describes a selection of case studies that present approaches for assessing the effectiveness of the CAAA in reducing acid deposition and improved methods for quantifying recent trends. Although there are several forms of deposition, the analyses herein focus on weekly or daily precipitation samples of sulfate concentrations. Similar to previous work in assessing trends in ozone concentrations, these analyses focus on estimating trends in data adjusted for the influences of meteorology.

In the United States, previous deployment of current federal, state, and private monitoring networks was primarily a response to the concern that acid deposition was causing adverse effects in lakes, streams, and forests, the extent and magnitude of which were not well characterized. The networks were established to answer several major questions: Where is the problem (patterns)? How severe is it (status)? Is it changing (trends)? Wet deposition is the most extensively measured component of deposition. Brief descriptions of existing monitoring networks are given in the next section.

The reader is referred to the *web companion* for specific data sets and software that are related to the case studies in this chapter.

2 Monitoring Data

2.1 Case Study I

The Multistate Atmospheric Power Production Pollution Study (MAP3S) monitoring network was designed to address research tasks covering pollutant emissions to the atmosphere, pollutant properties, distribution, transport, transformation, and removal processes. The operation of this network was one of several responses to the need for a better understanding of pollutant scavenging processes. The MAP3S monitoring domain is in the northeastern quadrant of the United States nine sites were located in areas not monitored by other networks and unaffected by local sources of pollution and urban influences. Data representative of the northeastern United States were sought (MAP3S, 1979). Of the nine MAP3S precipitation chemistry sites, four stations occupy an approximate north–south line from Whiteface Mountain, NY to Charlottesville, VA. Other stations are located at the Oak Ridge National Laboratory in Tennessee and at Lewes, DE on the Atlantic coast. Precipitation samples are collected nominally on a daily basis. Early in the operation of the network, samples were collected on the basis of a precipitation event.

2.2 Case Study II

The National Atmospheric Deposition Program/National Trends Network (or NADP/NTN) is the largest wet deposition network currently operating in the United States. The network is a cooperative monitoring effort involving many local, state, and federal agencies (NADP 1995). Its primary purpose is to establish a long-term acid deposition database that allows for accurate and precise prediction of broad spatial patterns and estimates of temporal changes. The network began operating in 1978 with 22 sites. As of 1995, there are approximately 200 sites in operation. Sites are selected to represent major terrestrial, aquatic, agricultural, and physiographic rural areas. All sites are well separated from urban and industrial regions and from large point sources, particularly power plants. Cumulative, weekly samples of precipitation are collected at each site and sent for analysis to the Central Analytical Laboratory at the Illinois State Water Survey laboratory. Samples are analyzed to determine the following ionic constituents of precipitation:

$$SO_4^{-2}, NO_3^-, Cl^-, PO_4^{-3}, Na^+, K^+, Ca^{+2}, Mg^{+2}, NH_4^+, \text{ and pH}.$$

Wet concentrations summaries are calculated as precipitation-weighted means (PWM) where for the period under consideration, the PWM is calculated from valid weekly concentration samples weighted by the rain-gauged measured precipitation. Wet deposition is calculated as the product of the PWM and the total measured precipitation. Further monitoring and analytical details can be found in NADP (1995).

The Canadian Air and Precipitation Monitoring Network (CAPMoN) is sponsored by the Atmospheric Environment Service of Environment Canada. CapMoN is a regional-scale deposition and air monitoring network that was implemented in 1983 by establishing 17 sites. As of 1995, 23 sites are in operation. The chemical constituents of primary concern to the network are the major ions in daily samples of precipitation and ambient sulfur and nitrogen compounds (Vet et al. 1986). Similar to the NADP/NTN process, sites were selected to minimize the effects of local sources of pollution.

The Acid Precipitation in Ontario Study deposition monitoring program operates a cumulative 28-day network (APIOS-C, where C designates cumulative sampling). Between 35 and 38 stations performed precipitation chemistry and air-quality sampling on a 28-day basis from 1981 to 1988 (Chan et al. 1982). APIOS-C remains in operation, but with a reduced number of sites. Similar to other deposition monitoring networks, monitoring locations were selected to avoid local and area sources of pollution that could prevent the site from being representative of its region.

2.3 Additional Ongoing Monitoring

The U.S. Environmental Protection Agency established the Clean Air Status and Trends Network (CASTNet) in 1987. CASTNet (U.S. EPA 1995) monitors wet and dry deposition of sulfur and nitrogen compounds. Dry deposition monitoring is conducted at 52 sites in the United States, most of which are located in the east. In cooperation with the National Park Service, dry deposition monitoring is conducted at 14 pristine sites in the western United States. Monitoring locations were selected according to strict siting criteria, so as to avoid undue influence from point sources, area sources, and local activities (e.g., agriculture). Dry deposition is extremely difficult to measure directly and has existed as a research component of the network since its inception. A number of investigators (Hicks et al. 1995; Wesely and Lesht 1988) have shown that dry deposition can be reasonably inferred by coupling air concentration data with routine meteorological measurements. Dry deposition is calculated as the product of ambient air concentration and a modeled deposition velocity that is itself a function of terrain and site meteorological data. Wet deposition information is collected at 20 sites to supplement NADP/NTN monitoring. Most of the CASTNet sites are located at or near sites where other agencies are conducting ecological or atmospheric research. The network was not designed to meet rigorous statistical design objectives, vis-á-vis estimating trends in sulfur or nitrogen deposition at a specific confidence level, because of unknown accuracy and interpolation errors in the estimation of dry deposition

Originally, its measurement was begun to establish its relative importance to wet deposition. It is important to note that sensitive ecosystems respond to the *total* deposition (wet plus dry) of a chemical species, not just wet or dry deposition separately.

3 Case Studies

3.1 Gamma Model for Trend Estimation

The estimation and associated significance testing of trends has been the subject of many investigations since the implementation of national precipitation monitoring networks in the late 1970s (Sisterson et al. 1990). Since concentrations (a PWM) mirror emission levels more directly than deposition (PWM multiplied by total measured precipitation), most investigations focus on discerning trends in concentration data. Because wet concentrations are known to depend on meteorological conditions, regression-type models (Baier and Cohn 1993; Sirois 1993; Lynch et al. 1995; Holland et al. 1995) have been used to remove variability caused by fluctuating weather at individual monitoring sites. Some investigations (UAPSP 1993; U.S. EPA 1994) have applied nonparametric methods (Sen 1968; Hirsch et al. 1982; Hirsch and Slack 1984), developed for the estimation of trends in water quality, to wet deposition data. Several analyses (Stein 1985; Egbert and Lettenmaier 1986; Le and Petkau 1988; Oehlert 1993) have modeled the space-time structure of wet deposition data collected at multiple monitoring sites to estimate trends in acid deposition.

Styer (1994a) applied gamma regression models to estimate site-specific trends in deposition data adjusted for the effects of season and meteorology. This approach assumed a constant coefficient of variation in the data (i.e., the standard deviation is proportional to the mean or, equivalently, a constant shape parameter in the gamma distribution). Many previous analyses applied a logarithmic transformation to the data and used statistical models that assumed normality of the transformed observations. Gamma models offer several advantages for regression modeling of such data: (1) The original units of measurement are preserved, obviating the need for back transformations; (2) the gamma distribution is appropriate for modeling deposition data that frequently exhibit positive, right-skewed frequency distributions; and (3) the gamma distribution overcomes a limitation of the logarithmic transformation that tends to produce skewed distributions of deposition calculated over monthly and longer time periods.

A gamma regression model was fit to daily sulfate concentration monitoring data observed at four sites in the Multistage Atmospheric Power Product Pollutions Study (MAP3S) network located in Whiteface Mountain, NY, Ithaca, NY, College Station, PA, and Charlottesville, VA. These data were collected on a daily basis during periods of precipitation over a 12-year span, 1976–1988. Based on previous studies, the regression model included meteorological covariates measuring temperature and wind conditions. The wind and temperature data used were from the National Meteorological Center gridded upper-air data. For each MAP3S site analyzed, 850 mb temperature and wind measurements from the closest grid point were used.

Daily sulfate concentration on day i was assumed to be gamma-distributed with expectation

$$\mu_i = \exp\left(\alpha_1 + \sum_{m=2}^{12} \alpha_m M_{mi} + \beta_p \log P_i + \beta_u U_i + \beta_v V_i + \beta_i T_i + \beta_d D_i + \beta_y Y_i\right), \quad (3.1)$$

where for the ith precipitation event, M_{mi} is an indicator variable for the month in which the precipitation event begins, P_i is the total precipitation amount in millimeters, U_i and V_i are the zonal and meridional components of wind, respectively, T_i is temperature in degrees Celsius, D_i is the number of days since the last precipitation event, and Y_i is the index variable for year. In particular, T_i was calculated as the daily temperature for the ith precipitation event minus the average daily temperature for the month during which the ith precipitation event began. Temperature deviations were expected to be positively associated with concentrations of sulfate, even after adjusting for season and wind direction. Warmer weather conditions led to increased photochemical oxidants for sulfur. The index variable for year was included to provide an estimate of long-term trend, and the number of days since the last precipitation event (D) was included to model the cleansing effect of precipitation. In the S-PLUS programming environment, the glm function can be used to fit this gamma model with a logarithmic link function. A link function describes how the mean depends on linear predictors; here,

$$g(\mu) = \exp\{\beta' x\}$$

or, equivalently,

$$\log(\mu) = \beta' x,$$

where $\beta' x$ is the linear predictor of the regression model.

Reflecting the keen scientific interest in this work, Styer (1994b) extended (3.1) to evaluate the effect of the temporal scale of the monitoring data on the estimation of long-term trends in sulfate concentrations. An assessment was performed on models that fit daily concentrations to daily meteorological covariates, weekly concentrations to weekly summaries of meteorology, and weekly concentrations to daily meteorological covariates at the four MAP3S monitoring sites examined in Styer (1994a). The full specification of the daily-daily model was given by (3.1). For the weekly-weekly model, it was necessary to create weekly measures of the meteorological covariates. Total weekly precipitation, mean weekly temperature, and wind measurements corresponding to the day of the week with maximum precipitation were used. Day count was defined as the number of days separating the first day of the week with rain from the previous rain day. One additional variable was included that represents the number of days in the week with rain.

Under the parameterization that is used for generalized linear models, the required assumption of a constant coefficient of variation is not necessarily valid with these data; indeed, the sum of the gamma distributed variables are not necessarily gamma distributed. Thus, if the daily concentrations are assumed gamma distributed, the weekly concentrations are not and this causes a slight model misspecification for the weekly data. Standard frequency diagnostics indicated that the gamma distribution was appropriate for the weekly data, however. The weekly-daily model was more complex in that it required the summation of daily summaries of meteorology with precipitation during a given week. Specifically, suppose the weekly sulfate concentrations are gamma distributed with expectation

$$\mu_j = \sum_{i=1}^{I_j} \exp(\beta' x_{ij}),$$

where I_j is the number of days with rain in the jth week, and daily measures of meteorology are included in x_{ij}, consisting of meteorological covariates for the ith day of rain during the jth week, including the seasonal and long-term trend model components. This model was fit by maximizing the associated likelihood directly because of the nonlinearity caused by using the logarithmic link function, making it impossible to use standard estimating routines for generalized linear models in the S statistical package.

Styer's (1994a) gamma regression model appeared to fit the data well, and her modeling assumptions were found to be valid. Jackknifed standard errors for the yearly trend term were calculated after positive autocorrelation was discovered in the residuals for each of the four sites. Sulfate concentrations were found to decrease significantly at each of the four sites, and trend estimates ranged from −1.1% to 3.5% per year across the sites. It was concluded that gamma models provide a highly flexible fit to these data and allow an easy interpretation of the results, preserving the original scale of the data. All three models of temporal aggregation were fit to the data, with subsequent comparison of the empirical estimates of the model parameters and associated standard errors. The results indicated that there can be small differences in the estimates of trend for the three temporal scales, as well as a loss of precision of these estimates in the aggregation. However, jackknifed estimates of the standard errors of the differences in the parameter estimates showed that there was no significant difference in the estimation of long-term trends using weekly data.

3.2 Network Ability to Detect and Quantify Trends

This analysis addressed the fundamental question of whether existing deposition monitoring could reveal changes in sulfate concentrations due to legislated emission reductions. Knowledge of the capability of current monitoring to estimate reliably projected CAAA changes in deposition was needed before an assessment could be made of the need to augment or delete monitoring sites from existing networks. This section describes the application of Oehlert's (1993) model to estimate the probabilities of quantifying future model-based year-to-year changes in sulfate concentrations. The analysis was based on monthly sulfate concentrations for the years 1982–1986 and stations in NADP/NTN, CAPMoN, and APIOS-C networks. Specific data screening criteria are given by Oehlert (1995).

A spatiotemporal model developed by Oehlert (1993) was used to estimate changes in the mean level of annual sulfate concentration over a period of several years. The time trend was modeled as a simple linear trend, although the methodology did not require this shape. The model structure incorporated three types of noise which appear in wet deposition data: (1) sampling and analytical errors; (2) effects due to short-term meteorology, lasting up to a few months, that produce autocorrelation in the data; and (3) effects of long-term meteorology that can influence large-scale spatial patterns in deposition. Estimating the trend in a tiled pattern of rectangles was the primary objective of this analysis. For it, the size of each rectangle was arbitrarily chosen to be about 1400 km^2. This discretized method of modeling and displaying trend complements our ability to display spatial patterns of deposition for fixed time periods.

The following model structure was assumed. Existing networks have s monitoring stations, each with y years of data. Let

$$Y_i' = (Y_{i1}, ..., Y_{iy})$$

be a vector containing the annual sulfate concentration at site i, let

$$Y' = (Y_1', ..., Y_s')$$

be a vector of concentrations for all the stations, and let $j(i)$ indicate the rectangle in which station i occurs. Then, the model characterized the structure of each deposition series, Y_i, as

$$Y_i = \alpha_{j(i)}\mathbf{1} + \mathbf{t}\beta_{j(i)} + L + N_i + \delta_i\mathbf{1},$$

where $\alpha_{j(i)}$ is the expected concentration for the rectangle, $\mathbf{t}$ is a trend shape or linear trend of length y centered to have mean zero, $\beta_{j(i)}$ is the expected concentration slope for the rectangle, L is a long-term noise series common to all stations, N_i is a short-term station-specific noise series, and δ_i is a station-specific effect accounting for measurement errors and bias that may be induced by elevation and proximity to point sources. A log transformation of annual precipitation-weighted concentration was used to help stabilize the interannual variance of sulfate concentration. One novel feature of this model was the inclusion of the continental spatial-scale component (L), and the consequent difficulty in modeling this component. While the existence of this component is open to debate, there is general agreement on the existence of wide-scale long-term phenomena in precipitation data. Thus, wet deposition, and to a lesser extent, sulfate concentration in precipitation should inherit at least part of this large-scale spatial variation through their dependence on precipitation volume.

Ordinary least squares (OLS) was used to produce estimates of the mean concentration (α) and slope (β) for each station. To account for the correlated error structure across time and space, regression residuals were used to estimate the spatial and temporal covariance structure. This provided input toward estimating the variance of site and regional trend estimates. To estimate the trend for each rectangle, neighboring rectangles were assumed to have similar values, hence a discrete smoothness prior distribution could be applied to the slope to model this relationship. Regional trends and associated variances were obtained by averaging rectangle trends and variances within a specified region. Further details of the modeling approach are given by Oehlert (1993).

The Regional Acid Deposition Model (RADM) (Chang *et al.* 1990) was used to project a future deposition trend scenario across the CAAA emission reduction period (1994–2003) for the mid-Atlantic, southern, mid west, and northeast CASTNet contiguous geographic regions. These regions represent areas of similar topography, wet deposition, and climatic characteristics. The reduction for each region occurs in seven unequal steps at the years: 1994, 1995, 1996, 2000, 2001, 2002, and 2003. The total sulfate reductions for these regions are 29%, 30%, 30%, and 32%, respectively, by the year 2003. This scenario is motivated by the belief that not all of the emission controls will occur in the mandated years of 1995 and 2000 but will, in fact, occur near the mandated years. It was impossible to investigate the entire range of reduction scenarios that may be of interest, but this projection is sufficiently plausible to give some potential conclusions regarding the ability of existing networks to distinguish changes due to emission reductions.

Oehlert's (1993) model can be used to determine the probability that the estimated change for a geographical region is within a factor of an hypothesized change predicted by RADM. These calculations depend on the magnitude of the predicted future change and the variance of regional change. For this model, the variance depends on t, the spatial configuration of monitoring sites (number and locations), and several other parameters. The spatial covariance, S, is the variance of the short-term noise series, and the off-diagonal elements for two stations separated by a distance d are $a + be^{-cd}$. Unfortunately, the temporal covariance, D, of the long-term error series, is difficult to specify. Herein, D is modeled via an ARMA(1,1) correlation structure (Box and Jenkins 1976) with coefficients (0.3, 0.95) times the variance of the long-term noise estimated from the covariance between sites separated by large distances. The coefficients of the ARMA model are based on an analysis of historical precipitation records spanning the period 1837–1957. Although precipitation and deposition are not identical, it is assumed that long-term patterns in precipitation should induce long-term patterns in deposition. The temporal correlation of the short-term noise series, C, is a tridiagonal matrix with ones on the diagonal and with the sub and super diagonal nearly zero. How to fit these variance parameters to the data and how to apply a discrete, two-dimensional smoothing distribution to estimate β_j in each rectangle j is described by Oehlert (1993; 1995). Given all of these modeled inputs, alternative configurations of new sites can be evaluated by varying the locations of some number of new sites and seeing how the variance of regional trend changes. Therefore, the model can be used to locate optimally new stations, locate redundant stations, and estimate the statistical power of seeing future changes in deposition.

For network design purposes, the data were adjusted for the effects of precipitation and linear time trend prior to variance estimation. Precipitation-adjusted annual averages were formed by adjusting the monthly data for each station by a linear effect for precipitation volume, and forming annual values by taking volume-weighted means of the adjusted monthly values. This adjustment for precipitation produces a substantial reduction in the spatial covariance compared to the covariance of the unadjusted data. Then, these annual volume-weighted, precipitation-adjusted values are detrended across years. Residuals from this linear adjustment for time are used to compute the variance of a station trend. Our basis to detrend over time is based on the economic factors that changed over 1982–1986, and on the expected changes in deposition that these factors caused. If, in fact, emissions remained relatively constant over this period, then we have underestimated variability and, consequently, overestimated the power of trend detection and quantification.

Reliable detection of a decreasing trend in concentration of sulfate in precipitation should occur by 1996 in the eastern United States, when it is projected that sulfate will have decreased about 18% from pre-CAAA levels. The probability of correctly detecting the RADM decrease, or "power" (i.e., concluding that sulfate concentration levels are monotonically decreasing), is quite good, with all regions having a power exceeding 0.9 by the end of 1997. By 1998, the power is essentially 1.0. This result holds even if the variance of the continental-scale component is underestimated by a factor of 2. To address the more difficult question of quantifying the estimated decrease, we have investigated how likely the estimated decrease is to be within ±20% of the predicted RADM decrease. Quantification according to this standard is much more difficult than trend detection. For the eastern regions, expect a 90% chance of being within 20% of the true RADM projection by the end of 2001, at which time sulfate is predicted to have decreased 22% from pre-CAAA levels.

In times of reduced budgets for large monitoring networks, it is useful to consider where sites can be added for maximal improvement to the network, or deleted to have minimal effect on network capabilities. Although many design criteria are plausible, if we consider (1) the sum of the regional variances in the five eastern United States regions and (2) the sum of the 40 largest rectangle variances in the eastern United States, Oehlert's (1993) model can be used to gain some insight concerning where to add or delete sites. The first criterion is designed to minimize the average regional variance, while the second attempts to keep the largest local variance from getting too large. To minimize the computational effort, stations are chosen sequentially. Specifically, the station that least increases a given criterion is determined. Given that station, the second station that gives the smallest increase in that criterion is determined, and so on. This sequential selection will not, in general, find truly optimal sets of stations, but should find sets that are not too far from optimal. For instance, to optimally delete 10 sites, most deletions would occur in the Midwest under both criteria, where the spatial configuration of sites is fairly dense. For either case, sites are not deleted near boundary areas where estimation is more difficult. Compared to variances computed for all sites operating in 1987, both sets of deleted sites result in little increase in each criterion. Conversely, adding 10 sites (where rectangle centers are potential site locations) results in placing almost all sites along the U.S. east coast. When selecting for regional variances, all new stations are in the northeast and middle Atlantic regions; when selecting for rectangle variances, the stations are spread out along the entire east coast. For each criterion, variance sums are reduced by 10–15%.

These recommendations for deleting and adding sites assume that all remaining sites continue to operate and that the optimization criteria are appropriate for making a decision regarding augmenting or reducing the size of the network. A log transformation is used to emphasize the variance at sites with low sulfate concentrations relative to sites with high concentrations. If trends at sites with low concentrations are not important, other transformations should be considered.

Oehlert (1996) extended his 1995 network design analysis to address the question, Which 100 stations of the 195 NADP/NTN stations in the United States. could be deleted if we wanted the remaining network to have the smallest possible trend estimate variances? Two criteria were used that depend on the variance/covariance structure of the estimated rectangle trends. The first information criterion is simply the sum of the 616 rectangle trend estimate variances covering the United States. The second is based on dividing the United States into 11 regions and computing the average of the rectangle variances for each region, then summing these variances across the 11 regions. Several sets of monitoring stations were evaluated separately for two values of the smoothing parameter (100, 25). Larger values of this parameter imply that adjacent rectangle trend values are more similar, so that the trend surface is smoother. The stations sets included (1) all stations, (2) 100 stations deleted sequentially using the total regional variance criterion, (3) 100 stations deleted sequentially using the rectangle variance criterion, and (4) 100 sets of stations with 100 randomly chosen stations deleted in each.

For a smoothing parameter of 100, optimizing (i.e., sequentially deleting 100 sites) the total regional variance produced a 7% increase compared to the total regional variance for all sites. The stations selected using the rectangle criterion had a 17% increase. Randomly chosen sets produced totals with a mean increase of 48%. Results for a smoothing parameter of 25 had variances approximately twice as large than for a parameter of 100. For stations deleted with the regional criterion, there was a 10% increase, while stations deleted with the rectangle criterion showed a 25% increase. Random deletions had a mean increase of 90%. Overall, the total rectangle variance produced a uniform distribution of sites across the United States, while the total regional variance maintained sites in or near small regions. Neither criterion led to the deletion of sites in boundary areas, where estimation was more difficult. These computations suggest that there can be a substantial reduction in the number of monitoring stations without a large increase in the evaluation criteria. Random station deletion, however, can lead to large increases in these criteria. Based on the length of the data record and modeling assumptions, Oehlert (1996) discusses the limitations of these conclusions.

4 Future Research

Most models estimate trend as a constant percent change per year. Clearly, however, more sophisticated summaries may be needed if sudden changes in deposition occur due to implementation of the CAAAs. For example, it may be necessary to parameterize trend as a step-function, or as some nonlinear form. Indeed, estimated trends over short time periods may not be indicative of persistent change in deposition rates. As more data become available, future assessments of trend should become more informative. It is difficult to compare trend analyses that are based on different data completeness criteria, selection of sites, time periods of data, and statistical methodologies. Future work should attempt to quantify the effects of different choices of data inclusion and methodology on trend estimation.

Inclusion of additional meteorological covariates may allow for better adjustments to deposition data and improved network design analyses. Potential sources of these data include output from meteorological models and U.S. National Weather Service monitoring stations. The sequential searching algorithms applied above are adequate, but stochastic methods such as simulated annealing (Aarts and Korst 1989) should be investigated to find optimal sets of sites to consider for inclusion or deletion.

Acknowledgment

I wish to acknowledge the useful comments of Gary W. Oehlert in this work.

References

Aarts, E. and Korst, J. (1989). *Simulated Annealing and Boltzman Machines.* Wiley, New York.

Baier, W.G. and Cohn, T.A. (1993). Trend analysis of sulfate, nitrate, and pH data collected at National Atmospheric Deposition Program/National Trends Network Stations Between 1980 and 1991. U.S. Geological Survey Open File Report 93-56. U.S. Geological Survey, Washington, DC.

Box, G.E.P. and Jenkins, G.N. (1976). *Time Series Analysis: Forecasting and Control.* Holden-Day, San Francisco.

Chan, W.H., Orr, D.B., and Vet, R.J. (1982). APIOS, an overview: The cumulative wet/dry deposition network. Technical Report ARB-15-82-ARSP. Ontario Ministry of the Environment, Toronto, Canada.

Chang, J.S., Binkowski, F.S., Seaman, N., et al. (1990). The regional acid deposition model and engineering model, NAPAP SOS/T Report 4. *Acidic Deposition: State of Science and Technology*, Volume I. National Acid Precipitation Assessment Program, Washington, DC.

Egbert, G.D. and Lettenmaier, D.P. (1986). Stochastic modeling of space-time structure of atmospheric chemical deposition. *Water Resources Research* **22**, 165–179.

Hicks, B.B., Baldocchi, D.D., Hosker, R.P. Jr., Hutchison, B.A., McMillen, R.T., and Satterfield, L.C. (1995). On the use of monitored air concentrations to infer dry deposition. NOAA Technical Memorandum ERL ARL-141. National Oceanic and Atmospheric Administration, Washington, DC.

Hirsch, R.M. and Slack, J.R. (1984). A nonparametric trend test for seasonal data with serial dependence. *Water Resources Research* **20**, 729–732.

Hirsch, R.M., Slack, J.R., and Smith, R.A. (1982). Techniques of trend analysis for monthly water quality data. *Water Resources Research* **18**, 107–121.

Holland, D.M., Simmons, C., Smith, L., Cohn, T., Baier, G., Lynch, J., Grimm, J., Oehlert, G., and Lindberg, S. (1995). Long-term trends in NADP/NTN precipitation chemistry data: Results of different statistical analyses. *Journal of Water, Air, and Soil Pollution* **85**, 595–601.

Le, N.D. and Petkau, A.J. (1988). The variability of rainfall acidity revisited. *Canadian Journal of Statistics* **16**, 15–38.

Lynch, J.A., Grimm, J.W., and Bowersox, V.C. (1995). Trends in precipitation chemistry in the United States: A national perspective, 1980–1992. *Atmospheric Environment* **29**, 1231–1246.

Multistate Atmospheric Power Production Pollution Study (MAP3S) (1979). The MAP3S Precipitation Chemistry Network: Second periodic summary report (July 1977–June1978). Report PNL-2829. Pacific Northwest Laboratory, Richland, WA.

National Atmospheric Deposition Program (1995). Precipitation chemistry in the United States, 1995. *NADP/NTN Annual Data Summary*. National Atmospheric Deposition Program, Natural Resource Ecology Laboratory, Colorado State University, Fort Collins, CO.

Oehlert, G.W. (1993). Regional trends in sulfate wet deposition. *Journal of the American Statistical Association* **88**, 390–399.

Oehlert, G.W. (1995). The ability of wet deposition networks to detect temporal trends. *Environmetrics* **6**, 327–339.

Oehlert, G.W. (1996). Shrinking a wet deposition network. *Atmospheric Environment* **30**, 1347–1357.

Sen, P.K. (1968). Estimates of the regression coefficient based on Kendall's tau. *Journal of the American Statistical Association* **63**, 1379–1389.

Sirois, A. (1993). Temporal variation of sulfate and nitrate concentrations in precipitation in eastern North America: 1979–1990. *Atmospheric Environment* **27A**, 945–963.

Sisterson, D.L., Bowersox, B.C., Olsen, A.R., Meyers, J.P., Vong, R.J., Simpson, J.C., and Mohnen, V. (1990). Deposition monitoring: Methods and results. State of Science and Technology Report No. 6. National Acid Precipitation Assessment Program, Washington, DC.

Stein, M. (1985). *A Simple Model for Spatial-Temporal Processes with an Application to Estimation of Acid Deposition*. SIAM Institute for Mathematics and Society, Philadelphia.

Styer, P.E. (1994a). An illustration of the use of generalized linear models to measure long-term trends in the wet deposition of sulfate. Technical Report 18. National Institute of Statistical Sciences, Research Triangle Park, NC.

Styer, P.E. (1994b). The effect of temporal aggregation in models to estimate trends in sulfate deposition. Technical Report 19. National Institute of Statistical Sciences, Research Triangle Park, NC.

U.S. Environmental Protection Agency/Office of Research and Development (1994). A Clean Air Act exposure and effects assessment 1993–1994, a prototype biennial assessment. Technical Report EPA/600/X-94/020. Environmental Protection Agency, Washington, DC.

U.S. Environmental Protection Agency/Office of Research and Development (1995). CASTNet National Dry Deposition Network. Technical Report EPA/600/R-95/086. Environmental Protection Agency, Washington, DC.

Utility Acid Precipitation Study Program (1993). Analysis of variability of UAPSP precipitation chemistry measurements, Final Report.

Vet, R.J., Sukloff, W.B., Still, M.E., and Gilbert, R. (1986). CAPMoN precipitation chemistry data summary: 1983–1984. Report AQRB-86-001-M. Atmospheric Environment Service, Downsview, Canada.

Wesely, M.L. and Lesht, B.M. (1988). Comparison of RADM dry deposition algorithm with a site-specific method for inferring dry deposition. *Journal of Water, Air, and Soil Pollution* **44**, 273–293.

Airborne Particles and Mortality

Richard L. Smith
University of North Carolina at Chapel Hill

Jerry M. Davis
North Carolina State University

Paul Speckman
University of Missouri

1 Introduction

In London in December 1952, a combination of adverse weather conditions and soot in the atmosphere produced one of the most lethal smogs in history, resulting in thousands of deaths. Similar events had occurred earlier in the Meuse Valley in Belgium in 1930, and in Donora, Pennsylvania in 1948. They focused attention on the need to avoid excessive levels of soot in the atmosphere. In Britain, the Clean Air Act of 1956, which, among other things, placed severe restrictions on the use of coal for home heating, resulted in a tenfold reduction in soot levels in London by the late 1960s. In the United States, the creation of the Environmental Protection Agency (EPA) and the passing of successive Clean Air Acts by Congress led to the enforcement of air pollution standards which have laid down strict controls on levels of ozone and particulate matter in the atmosphere. Nevertheless, a series of studies since the late 1980s have suggested that current standards are by no means strict enough. *The New York Times* (July 19, 1993) reported that up to 60,000 people a year are dying prematurely in the United States as a result of particulate matter pollution which for the most part lies within current EPA standards. A similar calculation in the British science magazine *The New Scientist* (March 12, 1994) concluded that 10,000 people die prematurely each year in England and Wales, as a result of atmospheric particulates. Such reports, backed up by many papers in the scientific literature, naturally led to calls for action, and in November 1996, the EPA issued new draft standards for ozone and particulate matter. These standards were attacked by industrial and other groups claiming that they would be expensive to implement and that the scientific case for new standards was very far from proven. In July 1997, the new standards were confirmed by the EPA, although with a delayed timetable for their implementation to allow for additional scientific research.

At the core of the scientific debate is a series of papers (Schwartz and Marcus 1990; Schwartz and Dockery 1992a, 1992b; Pope *et al.* 1992; Schwartz 1993; among others) which examined the relation between daily deaths and particulate levels in a number of cities. These studies were all essentially statistical analyses (reviewed in more detail in Section 2) in which some variable representing daily mortality was regressed on a number of covariates representing both air pollution and meteorology, with suitable adjustments to allow for seasonality, long-term trends, and serial correlation. In all cases, it was claimed, there is a statistically significant effect due to particulates after adjusting for other known causes of variation.

The present case study focuses on the issue at the center of this whole debate: the claim that there is a direct and quantifiable relationship between particles in the atmosphere and nonaccidental deaths in the human population, in particular among those aged 65 and over. Although there have been other studies related to different aspects of the problem, for example, hospital admissions, it is those studies focused on mortality which have produced the strongest effect and which have been at the forefront of calls for a new particulate matter standard. What has emerged is a delicate situation deserving close scrutiny. There is a lack of adequate biochemical explanation for the supposed short-term effect; the claimed effects are very small; the data are inadequate to assess exposure on individuals. The net effect of these circumstances leads to an inordinate influence of model (notably variable) selection in determining conclusions, calling into question, among other things, the validity of the asserted statistical significance levels.

Previous NISS research (Styer *et al.* 1995) has covered data from Cook County, Illinois (city of Chicago) and Salt Lake County, Utah (Salt Lake City). These are reviewed briefly in Section 5, along with some reanalyses of the Chicago data. The present chapter is primarily concerned with a data set from Birmingham, Alabama, originally treated by Schwartz (1993), which brings the issues in the previous paragraph into particular focus. The data consist of daily death counts, daily data on particulates, and various meteorological variables which form additional explanatory and possibly confounding variables. We consider whether the influence of particulates on mortality is statistically significant when various other factors are taken into account. We find that the claimed particulates–mortality relationship is highly sensitive to various arbitrary factors in the analysis, such as precisely which combination of lagged particulate values is taken to represent the "exposure measure." We also find that the effect in Birmingham is nonlinear, most of it being concentrated in the range above the current standard for annual average particulate levels — in other words, within the range that is already regulated by EPA standards. When these conclusions are considered alongside those reached from other data sets, it emerges that similar issues arise throughout these studies. Our results do not eliminate the possibility that there may be a genuine relationship which persists even at low levels of particulates, but they do demonstrate the very real difficulties which exist in drawing such conclusions from observational studies of this nature, in the absence of a clear scientific explanation.

2 Statistical Studies of Particles and Mortality

One of the first papers to consider the particulates–mortality relationship carefully was Schwartz and Marcus (1990). In that paper, the authors addressed a number of the difficulties involved in inferring a causal relationship from the available data. Among these are

- the effect of autocorrelation,

- the influence of long-term trends,

- the possible existence of a threshold level of particulates, below which there is no observable effect,

- whether the particulates effect is confounded with the weather,

- whether the effect due to particulates can be separated from that due to other forms of air pollution.

They applied their analysis to data from London over 1958–1972 and found a strong relationship between particulates and mortality over the whole range of levels of particulates. However, the change in mean level of the measured variable "British Smoke" was dramatic: a tenfold reduction over the period of the study. It is not surprising that such a strong long-term change in particulate levels should have resulted in a corresponding decrease in deaths. The difficult question, raised by the authors themselves and the focus of all the more recent discussion, is whether the relationship persists with the much lower levels of particulates in urban environments today.

Subsequent studies included analyses of Philadelphia (Schwartz and Dockery 1992a; Moolgavkar et al. 1995), Steubenville, OH (Schwartz and Dockery 1992b), Utah Valley (city of Provo — Pope et al. 1992), Birmingham (Schwartz 1993), and Chicago (Styer et al. 1995). Among the most comprehensive reports are two studies commissioned by the Health Effects Institute (Samet et al. 1995, 1997) which reviewed all the Schwartz studies as well as some independent data sets.

Although these papers all claimed to find a relationship between particulates and mortality, they differed in many important details of the analysis. Among these was the way they defined measures of particulates, either TSP (total suspended particulates) or PM_{10} (particulate matter of aerodynamic diameter less than or equal to 10 μm). They also differed in the way they took account of meteorological variables (mostly derived from temperature and humidity), how they handled other pollutants such as ozone and sulfur dioxide, and how they treated the seasonal effect. Models varied from one study to another, as did the conclusions. Some studies drew attention to features which call into question a simple causal interpretation, such as seasonality of the effect (Styer et al. 1995) or the collinearities of PM_{10} with other atmospheric pollutants (Moolgavkar et al. 1995; Samet et al. 1997). None of these studies was actually based on PM$_{2.5}$ (particulate matter of aerodynamic diameter less than or equal to 2.5 μm) although this is the variable on which the new EPA standards have been based, since it is apparently widely believed that it is the very small particles which penetrate furthest into the lungs and which therefore cause the greatest damage. It has also been suggested that PM$_{2.5}$ is more likely than PM$_{10}$ to permeate indoors, making an association between ambient levels and mortality more plausible than if this was not the case.

Other analyses have been of a prospective nature in which a large group of subjects is monitored for a number of years (Dockery et al. 1993; Pope et al. 1995). These studies have played a substantial role in the public debate; although from a statistical point of view, they are concerned with completely different issues from the analyses based on daily death counts — for example, the fact that their conclusions are based on comparisons *between* rather than *within* cities raises questions of ecological biases. We are not in a position to evaluate these studies and therefore do not consider them any further in the present chapter.

The reader is referred to the *web companion* for specific data sets and software that are related to the case studies in this chapter.

3 An Example: Data from Birmingham, Alabama

3.1 Summary of Available Data

The main data set used in the study of Birmingham is very similar to those used in previous studies, and consists of

- daily deaths from nonaccidental causes among the population of Birmingham aged 65 and over,

- daily PM$_{10}$ readings (μg/m^3) from monitoring stations within the city; where more than one reading was available for a given day, the average was taken,

- daily maximum temperature in °C (tmax),

- daily minimum temperature in °C (tmin),

- daily mean specific humidity in g/kg (mnsh),

- daily mean dew-point temperature in °C (dptp).

These variables are extracted from a larger data set, fully documented in Smith et al. (1998). Earlier experience suggests that the particulates effect is strongest in the 65 and over age group, so we concentrate on that group here. The PM$_{10}$ data come from the EPA's aerometric database, the mortality

Variable	mean	SD	min	10%	25%	50%	75%	90%	max
tmax	23.386	8.612	−3.3	11.1	17.2	24.4	31.1	33.3	38.3
tmin	10.654	8.913	−12.2	−2.2	3.3	11.1	19.4	21.1	25.6
mnsh	9.311	4.895	0.8	3.1	4.9	8.7	14.0	16.2	18.3
pm10	47.227	23.768	8.0	21.0	29.0	44.0	59.3	79.0	163.0
pmmean	46.888	18.960	13.0	24.2	33.0	44.7	57.8	72.0	137.1
dptp	10.538	9.317	−19.1	−2.9	3.3	11.8	19.1	21.5	23.5
Deaths	15.055	4.252	3.0	10.0	12.0	15.0	18.0	21.0	32.0

TABLE 1. Selected summary statistics.

data were obtained from the National Center for Health Statistics, and the meteorological data came from the National Climatic Data Center in Asheville, NC.

Summary statistics for all these variables are given in Table 1. Both the mean and the median of daily PM_{10} values are slightly below the EPA standard for the long-term mean, whereas is 50 $\mu g/m^3$, the data set contains only one value above the permitted daily maximum of 150. Also included in this table is "pmmean," defined as a three-day average of PM_{10}. This will be used later on as a regressor in preference to single-day PM_{10} values. The data cover the years 1985–1988, although daily PM_{10} data are only available from August 3, 1985, so the detailed analysis is confined to that period. Of the 1247 days thus covered, 99 had no PM_{10} reading and so are treated as missing days. In calculating pmmean, however, if only one or two days' data were available in any three-day period, then these were averaged to produce pmmean. By this definition, only three days out of the 1247 did not have a pmmean value. The summary statistics for deaths and for PM_{10} are a little different from the corresponding values in Table 1 of Schwartz (1993), although since in the discussion to follow we are able to produce similar results to Schwartz when we make similar analyses, it does not appear that this is a major reason for the differences in conclusion which we eventually reach.

Apart from Schwartz (1993), data from Birmingham have been analyzed by Samet et al. (1995), who essentially confirmed Schwartz's numerical results but without considering model selection aspects, by Roth and Li (1996), who focused particularly on model selection but from a different point of view to the one considered here, and by Clyde and DeSimone-Sasinowska (1997), who treated model selection via an automatic Bayesian algorithm.

3.2 Statistical Modeling Strategy

Most models in the analysis are of the form of a standard linear regression

$$y_t = \sum_j \beta_j x_{jt} + \varepsilon_t \qquad (6.1)$$

in which y_t represents deaths on day t or some transformation thereof, the $\{x_{jt}\}$ variables are known covariates, and $\{\epsilon_t\}$ are random errors. We have also considered Poisson regression models in which the distribution of y_t is assumed to be Poisson, the mean following a relation of form

$$\log \mathrm{E}\{y_t\} = \sum_j \beta_j x_{jt}. \tag{6.2}$$

In (6.1), y_t is usually chosen to be either the logarithm or the square root of daily nonaccidental deaths in the population aged 65 and over. The logarithmic transformation has the advantage of interpretability (each coefficient may be expressed as a percentage increase or decrease of mortality associated with the corresponding covariate) and compatibility with the Poisson regression approach (6.2). On the other hand, the square root transformation is variance stabilizing when the true distribution is Poisson, and in the present data set, this turns out to be the more significant consideration. The model (6.2) may be fitted by numerical maximum likelihood, which has been implemented directly in Fortran, or using the generalized linear modeling (glm) routines in S-PLUS. Model (6.1) may, of course, be fitted by ordinary least squares, which has the advantage for model exploration purposes of not requiring iteration and, therefore, being faster.

The independent variables $\{x_{jt}\}$ are of three types: smooth terms representing long-term trends, meteorological variables, and air pollution. We now consider each of these in turn.

Modeling Trend and Seasonality

In all cases that have been studied, data on long-term mortality rates show substantial seasonal fluctuations and long-term trends over and above anything that can be explained by either air pollution or meteorology. It is, therefore, essential to model this dependence. For Birmingham, Figure 1 shows weekly mortality totals for the four years 1985–1988, with a *loess* curve (solid curve) to represent the smoothed underlying trend, and a B-spline representation discussed below (dashed curve). The vertical lines represent ends of years. It can be seen that there is a strong seasonal effect, but it is irregular — for instance, peak deaths during the winter of 1986–87 occurred in early January but for 1985–86 and 1987-88, the peak was in late February. Such variability could be due to epidemics such as influenza. In Schwartz (1993), the trend was modeled as a periodic function based on a two-year cycle through a sine-cosine representation. For this particular data set, this idea appears to work quite well, but as a general modeling strategy, it seems more satisfactory to apply general methods of nonparametric function fitting, which have already been reviewed in Chapters 2 and 3. One particular method which is widely favored is to represent an arbitrary smooth function using splines, which are piecewise polynomial functions with smooth connections between one segment

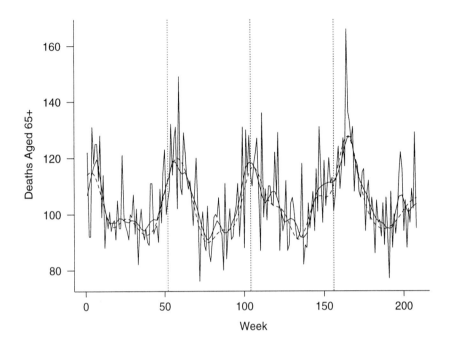

FIGURE 1. Plot of weekly total nonaccidental deaths aged 65 and over in Birmingham, together with fitted loess curve (obtained in S-PLUS with f=.05; solid curve) and B-spline fit with 20 df (dashed curve).

and another. A convenient way to achieve this is to use cubic B-splines (Green and Silverman 1994, pp. 157–158), in which the trend from time 0 to time T is represented as

$$f(t) = \alpha_0 + \sum_{k=1}^{K-1} \alpha_k \delta_k(t), \ 0 \leq t \leq T, \qquad (6.3)$$

where $\delta_1(\cdot), ..., \delta_K(\cdot)$ are defined by

$$\delta_k(t) = B\left\{\frac{K}{T}(t - \tau_k)\right\} - \frac{1}{K}. \qquad (6.4)$$

Here, $B(\cdot)$ is the "B-spline basis function" formally defined by

$$B(x) = \begin{cases} 0, & x \leq -2, \\ (x+2)^3/6, & -2 \leq x \leq -1, \\ \{-3(x+1)^3 + 3(x+1)^2 + 3(x+1) + 1\}/6, & -1 \leq x \leq 0, \\ (3x^3 - 6x^2 + 4)/6, & 0 \leq x \leq 1, \\ (2-x)^3/6, & 1 \leq x \leq 2, \\ 0, & x \geq 2. \end{cases}$$

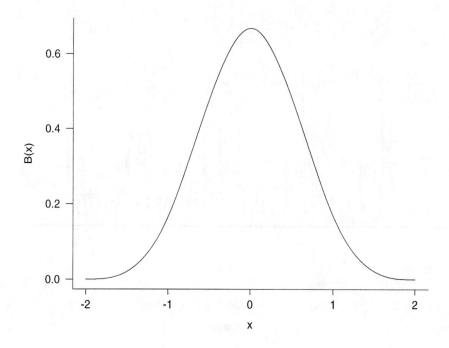

FIGURE 2. Cubic B-spline basis function.

A plot of the B-spline basis function is shown in Figure 2. It is a non-negative function, with continuous derivatives up to third order, and total integral 1. In (6.4), we take $\tau_k = Tk/K$ to be the kth "knot" and K to be the total number of knots. The knots have to be equally spaced for the formula (6.4) to work. The reason for subtracting $1/K$ in (6.4) is to center $f(t)$ around 0. The reason why the sum in (6.3) stops at $k = K - 1$ rather than $k = K$ is to avoid an indeterminacy. A similar though not identical representation may be generated in S-PLUS using the bs function. The reader is also referred to Chapter 3 and Appendix A for a more general discussion of splines and their computation.

The main methodological issue in this approach is the selection of K, the number of basis functions. Following exploratory work by Davis et al. (1996), we have chosen K to correspond to one "knot" every two months, which is used for the dashed curve in Figure 1 and all subsequent fits.

Meteorological Covariates

Meteorology is important because there are well-documented effects due to extreme meteorological conditions — in particular, very cold weather in the winter, and a combination of high heat and humidity in the summer — which could act as confounders with air pollution. However, there is also a danger of overfitting the meteorological effects, as could well occur if we included every conceivable meteorological variable in the regression analysis. For this reason, we include only a limited number of meteorological variables which are known to have an effect. The ones considered in this study are daily maximum temperature[1] (tmax), daily minimum temperature (tmin), and specific humidity (mnsh). Some authors have used dew-point temperature in place of specific humidity; our own studies suggest that the models based on specific humidity fit only slightly better than those based on dewpoint (dptp), but it hardly matters which one is chosen. On the other hand, it does seem important to include some humidity-based variable in the analysis. For a further discussion of these issues, see Appendix B of Styer et al. (1995).

Even when the variables are fixed, it is still important to consider which lags should be included, and whether the effects should be taken to be linear or nonlinear. It is often found that a two- or three-day lagged variable is a better predictor of mortality than the current day's value. Sometimes, the best predictor is obtained through a combination of current and lagged values, sometimes with opposite signs in the coefficients. This could be interpreted as meaning that a sudden change in the weather is associated with increased mortality. These effects have been allowed for by including up to four-day lagged values of meteorology in the models and using variable selection strategies to select the best model.

As far as nonlinear effects are concerned, one would expect to see increased mortality due to temperature at either end of the scale, and this is supported by the data. For example, Figure 3 shows a plot of daily mortality against daily maximum temperature lagged one day. Also shown on the plot is a *loess* smoothed curve. The apparent effect is of a decrease in mortality with increasing temperature across most of the range, but with a change of slope somewhere near 30°C. Most plots of mortality vs. temperature show this kind of relationship, although the change point is different in different cities. This suggests modeling the relationship with a function of form

$$f(t) = \beta_1 + \beta_2(t - t_0)_+ \qquad (6.5)$$

($x_+ = \max(x, 0)$) corresponding to a change in slope at $t = t_0$. This can be fitted by including $(t - t_0)_+$ as a term in the regression. In our analyses, t_0 has been chosen through initial plots such as Figure 3, rather than formally selected by the regression analyses. We use $t_0 = 30$ for Birmingham.

The other meteorological terms are taken to be linear except for the case of specific humidity, for which a quadratic term also proved to be statistically significant.

---

[1] Thoughout this chapter, it will be useful to refer to the regression variables using abbreviations. These strings are convenient because they are also correspond to the S-PLUS data set names.

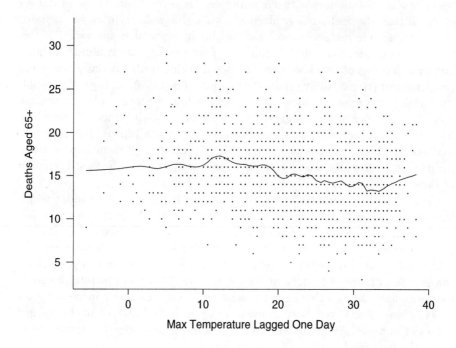

FIGURE 3. Plot of daily deaths vs. temperature lagged one day, along with fitted trend via *loess*.

Measures of Particulates

The other major issue to consider is precisely which variable or combination of variables best represents the particulates effect.

Previous authors have used a variety of variables. Schwartz and Dockery (1992a) used two-day averages of TSP in Philadelphia, Schwartz and Dockery (1992b) used a one-day lagged value of TSP in Steubenville, Pope et al. (1995) used five-day averages of PM_{10} in Utah Valley, Styer et al. (1995) used three-day averages of PM_{10} *including* the current day in Chicago (i.e., daily values with lags 0, 1, and 2 days), whereas Schwartz (1993) used three-day averages of PM_{10} *excluding* the current day (in other words, average of lags 1, 2, and 3 days) in his analysis of Birmingham. Clearly, the selection of this variable needs to be taken into account in assessing the statistical significance of the result. The strategy adopted here is to start with all five daily values (lags 0, 1, 2, 3, 4) and then use variable selection techniques to decide which variables to include. However, for comparison with previous analyses, three-day averages both including and excluding the current day will also be considered.

Another issue relevant to the debate over particulate matter standards is whether there exists a particulates "threshold" below which there is no discernible relationship between particulates and mortality. To examine such a question, it is necessary to consider nonlinear models for the particulates–mortality relationship, and in Sections 4.2 and 4.3, we shall consider different approaches toward this.

Modeling Strategy

Taking into account the various issues raised in the preceding subsections, we now outline a specific modeling strategy which is implemented in Section 4.

In all of our analyses, the objective is to model daily deaths as a function of meteorology, particulates, and trend. The meteorological variables were tmax and tmin, mnsh, mnsh^2, and PM$_{10}$, considered both for the current day (lag 0) as well as for 1-, 2-, 3-, and 4-day lags. This is denoted by, for example, tmax$_4$ to indicate tmax lagged 4 days. Note that tmax$_0$ means the current day's value. The lagged daily values for PM$_{10}$ are denoted pm$_0$,...,pm$_4$ and we also use pmmean as a notation for three-day averages of PM$_{10}$. Thus, pmmean$_0$ means the averages for lags 0, 1, and 2, while pmmean$_1$ means averages for lags 1, 2, and 3. We also included a variable tg30 and its lagged values, where tg30=(tmax–30)$_+$. This variable is included to permit a piecewise linear mortality–temperature relationship as discussed earlier.

In all analyses considered, a "best" model is first established using the meteorological variables plus 21 B-spline basis functions to represent the trend, and only then are the PM$_{10}$-based variables added to the model.

The basic model selection strategy is backward selection, starting with all variables in the model and dropping them one at a time, based primarily on t statistics, but considerable judgment was exercised in selecting which variables to drop so that the resulting model would be consistent with known risk factors. A decision was made not to drop tmax if tg30 (at the same lag) was still in the model, and similarly not to drop mnsh if mnsh^2 was in the model. Also, at the end, when PM$_{10}$ was added, variables which were identified during the meteorology-only phase of the model fitting were not dropped even if at that stage the t values no longer looked significant. This explains why some of the models reported later contain t values which are not significant. The coefficients of the B-spline representation of the trend were retained in all models. Obviously, there is a considerable amount of arbitrariness about the precise model selected, but the same strategy was repeated several times with different combinations of variables and a good level of consistency among the results, so that the final results do not appear to be highly sensitive to the exact model-selection strategy adopted.

Model	NLLH	df
(a) $tmax_1$, $tmax_4$, $tmin_3$, $tg30_1$, $tg30_4$	3481.965	26
(b) $tmax_1$, $tmax_4$, $tmin_3$, $tg30_1$, $tg30_4$, $mnsh_3$, $mnsh_4$	3479.841	28
(c) $tmax_1$, $tmax_4$, $tmin_3$, $tg30_1$, $tg30_4$, $dptp_3$, $dptp_4$	3478.662	28
(d) $tmax_4$, $tmin_3$, $mnsh_0$, $mnsh_2$, $tg30_4$, $mnsh_0^2$	3477.351	27

TABLE 2. Best meteorological models in four classes.

Variable	Coefficient	S.E.	t Value
$tmax_4$	.00837	.00465	1.80
$tmin_3$	−.00626	.00481	−1.30
$mnsh_0$	−.05489	.01776	−3.09
$tg30_4$	−.02008	.00829	−2.42
$mnsh_2$	−.03147	.01482	−2.12
$mnsh_0^2$	.00335	.00092	3.63

TABLE 3. Best linear regression model using temperature and humidity.

4 Results for Birmingham

4.1 Linear Least Squares and Poisson Regression

The first set of models considered were normal linear regression models based on square root of elderly (aged 65 and over) mortality as the dependent variable, after preliminary analysis confirmed that this was superior to using either no transformation or a logarithmic transformation. Poisson regression analyses are based on a logarithmic link function, as in (6.2).

For the linear regression of square root deaths, Table 2 shows negative log likelihood (NLLH) values for the final model selected in four classes: (a) using temperature variables only, (b) using temperature plus linear terms in specific humidity, (c) using temperature variables plus dew-point temperature, (d) using temperature variables plus linear and quadratic terms in specific humidity. The results show that (c) is better than (b), but (d) is better than either; compared with (a), (b) is not statistically significant ($P=.11$) but (c) is ($P=.03$). It is not possible to compute a P-value for (d) against any of the other three because the models are not nested, but the NLLH value and the t value for $mnsh_0^2$ (3.63; see Table 3) strongly suggest a significant result. Here, df is the number of estimated coefficients plus 21 (overall constant + 20 independent terms from the spline representation).

Table 3 shows the individual coefficients, standard errors, and t values for model (d) of Table 2.

Now, we consider what happens when PM_{10} is added to the model of Table 3. As noted already, both $pmmean_0$ and $pmmean_1$ have been used as "exposure

Variable	RR	95% Confidence Interval	P-Value
(i) pmmean$_0$	1.005	[0.996 , 1.015]	.26
(ii) pmmean$_1$	1.010	[1.000 , 1.019]	.048
(iii) pm$_0$,...,pm$_4$	1.008	[0.993 , 1.023]	.094
(iv) pm$_0$, pm$_1$, pm$_3$	1.007	[0.995 , 1.019]	.029

TABLE 4. Relative risks and confidence intervals due to a rise on 10 $\mu g/m^3$ of PM$_{10}$ in the model of Table 3.

measures" in previous analyses, and these are reflected in rows (i) and (ii) of Table 4. The contrast between the results, which are statistically significant for pmmean$_1$ but not for pmmean$_0$, motivated us to look in more detail at the influence of individual days, in row (iii). When all five variables pm$_0$,...,pm$_4$ are included in the model, the t values are respectively -1.8, 2.1, -1.0, 1.2, and 0.8. It was then decided to drop pm$_2$ and pm$_4$, which had the smallest t values, producing the results of row (iv), in which pm$_0$, pm$_1$, and pm$_3$ had respective t values of -1.9, 1.8, and 1.8. Thus, we have the seemingly paradoxical result that the coefficient for day 0 is negative, whereas those for days 1 and 3 are positive. The principal quantity tabulated in Table 4 is the relative risk (RR) corresponding to a rise of 10 $\mu g/m^3$ of PM$_{10}$, together with a 95% confidence interval, and the P-value for the PM$_{10}$ coefficients based on the NLLH. Note that the P-value for rows (iii) and (iv) refers to the combined significance of all the variables included; thus, for row (iv), the combined P-value is .029 (significant at .05) based on an F-test applied to the three variables pm$_0$, pm$_1$, and pm$_3$ together, but because the coefficients are of opposite sign, the confidence interval for the overall effect of particulate matter includes relative risk 1.

In choosing a reference level of PM$_{10}$ against which to calculate relative risk, we have selected 10 $\mu g/m^3$ because this is a reasonable guess at how much PM$_{10}$ levels might actually be reduced as a result of tighter regulations. Other authors have used 100 $\mu g/m^3$ as their reference level, but this seems misleading because we hardly ever actually observe a rise or fall of that amount.

As an example of the calculations involved in Table 4, consider row (ii). The pmmean$_1$ coefficient here is .00187 with standard error .00095, so a 95% confidence interval for the coefficient is [.00001, .00373]. The dependent variable is the square root of daily death count, whose mean is 15.055 (Table 1). Therefore, the estimated RR associated with a rise in PM$_{10}$ of 10 $\mu g/m^3$ is $(\sqrt{15.055} + .0187)^2/15.055 = 1.0097$. Repeating the same calculation for the upper and lower confidence limits produces the 95% confidence interval shown in Table 4. Admittedly, this is a somewhat rough calculation, but more sophisticated versions (e.g., for each day in the sample, increase the PM$_{10}$ level by 10, recompute the expected number of deaths based on the new PM$_{10}$ and all the other covariates, and then average over all days) produce almost exactly the same answers. The calculations for rows (iii) and (iv) are similar but based on a sum of coefficients and the corresponding standard error for the estimate of the sum.

These analyses have also been repeated using Poisson regression. The actual values of the regression coefficients are in this case quite different, because Poisson regression uses a logarithmic link function, whereas the preceding results have used a square root transformation. However, the qualitative results, including the RR calculations of Table 4, are almost identical if we use Poisson regression in place of least squares regression on square root deaths.

Two other issues which have been raised in previous analyses are overdispersion (in comparison with the variances implied by the Poisson model) and serial correlation. In this case, overdispersion is minimal — the variances of regression residuals are about 8% higher than a Poisson model would lead us to expect — and this appears to have minimal influence on the assessed significance of the PM_{10} or any other effects. Residuals have been tested for autocorrelation and found not significant.

The results may be summarized as follows. In comparing models based on temperature alone with those based on both temperature and humidity, it does appear that humidity should be included in the model, but this does not have a great influence on the significance of the PM_{10} component. The analysis confirms that when $pmmean_1$ is included, the coefficient is statistically significant. However, a more detailed breakdown into the effects of individual days presents a more complex picture, with a negative effect for lag 0 and positive effects for lags 1 and 3. It does not seem plausible that the negative coefficient for day 0 really represents a protective effect; it is much more likely to be due to correlations among the PM_{10}-based variables. Something similar was observed by Samet et al. (1997) for the effect of NO_2 in Philadelphia. At the same time, however, it seems disingenuous to omit one of the variables simply because it gives a coefficient of the wrong sign, as is implicitly being done when $pmmean_1$ is computed. The analyses in line (iv) of Table 4 represents the most realistic attempt to come up with an overall estimate of the effect of PM_{10}. Both it and the corresponding Poisson regression show a RR of around 1.007, but the result is not statistically significant, as indicated by the confidence interval. Note that the computation of the standard error has to take account of the fact that three coefficients are being estimated and this explains why the confidence interval is wider than in earlier computations.

4.2 Nonlinear Effects

A critical question for the whole particulates–mortality debate is whether there exists a threshold below which there is no discernible effect. This question can only be addressed through some form of nonlinear modeling. Schwartz (1993, Figure 6) estimated a smooth nonlinear curve for the PM_{10}–mortality relationship but did not calculate any confidence band. Other studies including

Range of x	Slope	S.E.	t Value
0–33	.00124	.00249	0.50
33–45	.00081	.00250	0.32
45–58	−.00060	.00223	−0.27
58–∞	.00186	.00099	1.88

TABLE 5. Slopes for each section of piecewise linear fit.

those of Samet et al. (1995, 1997) have used loess analyses within S-PLUS as a means of estimating nonlinear curves with confidence bands.

Two approaches are taken in this section, one based on a piecewise linear fit and the other on B-splines. An alternative based on GAM modeling, including loess fits, is in Section 4.3. Both models use pmmean$_1$ as the particulates-based covariate; this seems the only case which is even worth trying, since the others are not significant when fitted as linear models.

The piecewise linear approximation is based on adding the following four terms to the meteorological model: $(33-x)_+$, $(45-x)_+$, $(x-45)_+$, and $(x-58)_+$, where x is pmmean$_1$ and $y_+ = \max(y, 0)$. The change points 33, 45, and 58 were selected as the three quartiles of the distribution of pmmean$_1$. Defined in this way, the variables are centered around $x = 45$: In other words, what is being estimated is the *change* in mortality at a particular x compared with $x = 45$. Some centering of this nature is essential to make sense of the results.

When this model was fitted, the reduction in NLLH, compared with the model with just a single linear term in pmmean$_1$, was only 0.5 (3 df), which is certainly not statistically significant. Pointwise confidence bands may be obtained by noting that, for any given value of x, the estimated effect at x is just a linear combination of the four estimated coefficients and, so, its standard error may be obtained from the covariance matrix of the regression estimates. Because of the centering, the confidence bands are of width 0 at $x = 45$. A plot of the fitted piecewise linear curve, with pointwise 95% confidence bounds, based on the Poisson model fit, is shown in Figure 4. The confidence bounds suggest that the most statistically significant slope is in the upper part of the range of x and this is confirmed in Table 5 which shows slopes for the four sections of the plot.

A second approach has also been tried using B-splines. This has the advantage of producing smooth curves; otherwise, the results are similar to the piecewise linear representation. A B-spline representation was taken with four knots and log pmmean$_1$ as the independent variable, then transformed back to the original scale for plotting purposes. The results are shown in Figure 5. Once again the function is centered at $x = 45$, so that the confidence bands have width 0 at this point.

Figure 5a is very similar to Figure 6 in Schwartz (1993), including the dip around 50, but the confidence bands leave no doubt that this effect is spurious.

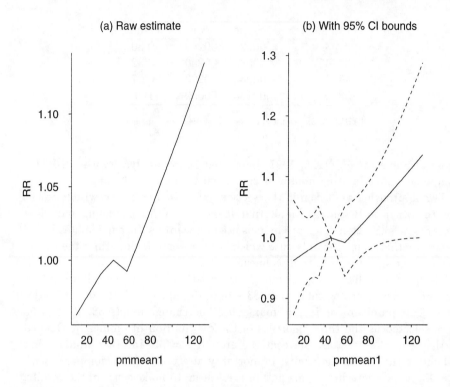

FIGURE 4. Piecewise linear fit of the relative risk due to pmmean$_1$, computed relative to pmmean$_1$=45. (a) Raw estimate; (b) raw estimate with pointwise 95% confidence bounds.

These plots do not directly answer the question of whether there is a threshold, but when the width of the confidence bands is taken into consideration, they suggest that it is only at the higher levels of PM$_{10}$ that there is anything close to a significant effect. An alternative approach to this question is to fit a single component of the form

$$(x-t)_+,$$

where x is the variable of interest (here, pmmean$_1$) and t is some threshold. In particular, $t = 0$ correspond to an ordinary linear term; whereas the limit as $t \to \infty$ represents no dependence on PM$_{10}$.

Taking $t = 0$ as a reference value, in Figure 6, the deviance (twice the difference in log likelihoods) is plotted as a function of t, along with confidence bands at ±3.84, corresponding to the 95% point of a χ_1^2 random variable. Values outside this confidence band may be interpreted as significantly worse than $t = 0$ if they are above the confidence band, or significantly better if they are below. As might be expected, very high values of t are significantly worse than $t = 0$ (although only just). This is just another way of saying that the linear trend based on pmmean$_1$ is significantly better than the model

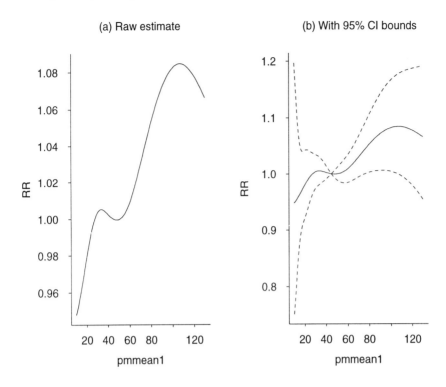

FIGURE 5. B-spline fit of the relative risk due to pmmean$_1$, computed relative to pmmean$_1$=45. (a) Raw estimate; (b) raw estimate with pointwise 95% confidence bounds.

with no PM$_{10}$ component. Note that, for this part of the discussion, we are ignoring completely all issues connected with the selection of pmmean$_1$ as the particulates-based variable. However, what is striking from Figure 6 is that most values of the deviance are either close to or below the value for $t = 0$, with a minimum at $t = 68$, which therefore becomes the "maximum likelihood estimate" of the threshold.

The fact that there are no values of t which are significantly better than 0 can be interpreted as saying that a null hypothesis $t = 0$ would be accepted against an alternative hypothesis based on any alternative value of t. In other words, the data do not reject a hypothesis that the relationship is linear across the entire range of PM$_{10}$ values. However, one could equally well take the null hypothesis to be $t = 50$, corresponding to the current EPA standard for the annual average. This, also, is accepted against any alternative value of t. Based on Figure 6, the data do not provide evidence to discriminate among any values of t below about $t = 80$.

The conclusion from this is that there is no evidence in the current data that lowering the current PM$_{10}$ standard would have any beneficial effect. The evidence in favor of a particulates–mortality relationship, such as it is, is based

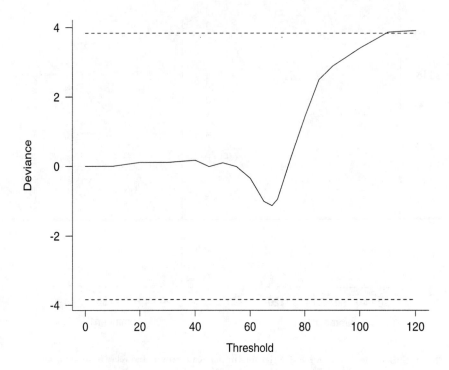

FIGURE 6. Plot of the deviance associated with the different threshold values, relative to that for threshold 0. The dashed lines are at the 95% confidence bands of ±3.84.

entirely on levels of particulates above the current standard for annual average PM_{10}.

4.3 Nonparametric Regression

In addition to the parametric approach taken in the analyses so far, a nonparametric approach has been undertaken using the framework of generalized additive models (GAMs; Hastie and Tibshirani 1990). The main advantage of the nonparametric approach is that it provides complete flexibility for exploring the relationship among a set of variables.

The basic additive model is defined by the equation

$$y_t = \sum_j f_j(x_{jt}) + \varepsilon_t, \tag{6.6}$$

where x_{jt} is the value of the jth covariate on the tth day. As in ordinary least squares, $E\{\varepsilon_t\} = 0$ and $Var\{\varepsilon_t\} = \sigma^2$. The f_j terms are arbitrary univariate functions with an f_j modeled for each covariate (Hastie and Tibshirani 1990;

Green and Silverman 1994). Note that in the linear case, $f_j(x_{jt})$ is equal to $\beta_j x_{jt}$; cf. (6.1).

It is possible to view the B-spline approach within this framework, but the results of the present section have taken the alternative *loess* approach (Chambers and Hastie, 1993). A brief description of this nonparametric regression method can be found in Chapter 3.

Model Selection

For this portion of the research, it was decided to fill in the missing 99 observations in the original PM_{10} data set; otherwise, a lag-four data set would have about 400 missing observations. (Note that the analyses of Sections 4.1 and 4.2 took the opposite approach, in which days with missing data were simply omitted from the analysis.) There were never more than three missing values in a row in the daily data set, and this occurred only three times. When one value was missing, the average of the adjacent two values was used. When two or three values were missing, a straight-line fit was applied. For example, if the PM_{10} data were 10 NA NA NA 50, with NA standing for a missing value, then the NAs (from left to right) would become 20, 30, and 40.

A GAM was initially developed for the meteorological effect using the S-PLUS function gam with backward selection and the AIC criterion. The selected variables constituted the basic meteorological model for all the subsequent GAM runs. The entire data set of 1247 observations was used to develop this model. The response variable was the square root of daily deaths in the 65+ age group. Each meteorological covariate was modeled as either a linear term or using the *loess* procedure. The default bandwidth (0.5) was selected after a complete range of bandwidths was considered for the covariates. The meteorological variables used were daily maximum temperature (tmax), daily mean temperature [mntp, defined as $\frac{1}{2}$(tmax+tmin)] and daily mean specific humidity (mnsh), each with lags up to four days. The trend term was modeled using B-splines with 20 degrees of freedom to correspond to earlier analysis.

Following the development of the basic meteorological model (hereafter, the basic model), the PM_{10} terms were added to this model and comparisons were run between the basic model and the basic model with PM_{10} terms added. Comparisons were done using an ANOVA procedure using the approximate F-test (Chambers and Hastie 1993). A statistically insignificant F statistic at the 0.05 level indicates that the addition of the PM_{10} term to the basic model has no significant explaining power with regard to the response variable. Three different categories of PM_{10} were added, as in Section 4.1, based on either $pmmean_0$ or $pmmean_1$ fitted on its own, or a combination of the five daily values $pm_0,...,pm_4$. However, a nonparametric F-test, applied to $pmmean_0$ and $pmmean_1$, indicated that neither variable was nonlinear in form.

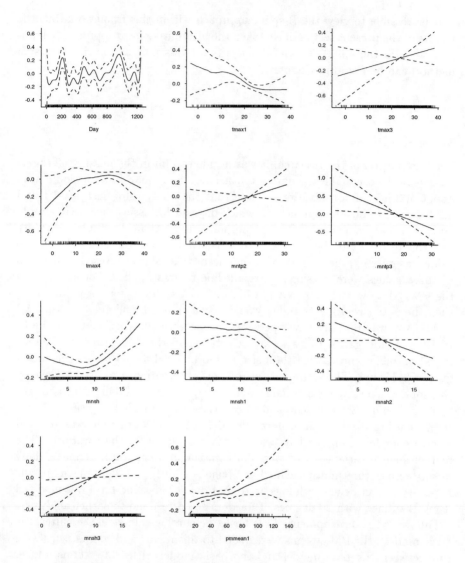

FIGURE 7. Plot of 11 fitted f_j functions in GAM model, with confidence bands.

GAM Results

The GAM stepwise algorithm identified the following variables as significant for the basic model: lo(tmax$_1$), tmax$_3$, lo(tmax$_4$), mntp$_2$, mntp$_3$, lo(mnsh), lo(mnsh$_1$), mnsh$_2$ and mnsh$_3$, together with the B-spline terms for the trend. Here, lo(...) indicates that a *loess* nonlinear function was fitted, while the remaining variables were all treated linearly. The dispersion parameter for the

Gaussian family was taken to be 0.272; i.e., a 9% increase compared with the value 0.25 that would be expected if the Poisson model were exact. This is consistent with estimates of overdispersion found in other analyses.

As in the earlier analyses, when the PM_{10} effect is represented either through $pmmean_0$ or through a matrix of one-day values, it is found not to be statistically significant. Only with $pmmean_1$ (P-value $= .026$) is a significant result obtained.

A further analysis was performed using pm_0, pm_1, and pm_3 as particulate matter variables to give results comparable with those in Section 4.1. An initial fit again showed a negative coefficient for pm_0 and positive coefficients for pm_1 and pm_3, with t values of -1.05, 2.25, and 1.19, respectively. In further analysis, first pm_0 and then pm_3 were dropped from the model, leaving pm_1 as the only significant variable, for which the coefficient was $.00159$ and standard error $.00079$ (P-value $.= 046$). Thus, the results differ somewhat from the earlier ones in that with this version of the analysis, the negative coefficient for pm_0 is not statistically significant. The conclusion from this analysis is that only pm_1 is significant, and then only just so.

Once again, the uncertainties associated with variable selection cast doubt on whether any of the PM_{10}-based variables is really significant. In Figure 7, for the model based on $pmmean_1$, the 11 fitted $f_j(x_j)$ functions are plotted, along with their pointwise 95% confidence interval bounds.

5 Comparisons with Other Cities

The first NISS paper on particulates and mortality was Styer et al. (1995). This paper used two data sets, one taken from Chicago (Cook County, Illinois) and the other from Salt Lake County, Utah. The data were of similar structure to those for Birmingham. The main statistical analysis technique was Poisson regression, using equation (6.2) to relate daily death y_t to covariates $\{x_{jt}\}$ based on trends and seasonal effects, meteorology, and, finally, particulates. This was guided by seasonal and month-by-month analyses using the semiparametric method deriving originally from Sacks et al. (1989), which is described in Chapter 2. The advantage of the latter, in this context, was that it provides a semiautomatic method for deciding which variables should be included in the model and whether they should be treated linearly or nonlinearly — if the latter, the method gives smoothed estimates of the nonlinear effect.

A major concern of Styer et al. was seasonality. In the analysis of Birmingham reported in Sections 3 and 4, seasonality was taken into account in the additive trend modeled via B-splines, but the regression coefficients of meteorological and particulate variables were not taken to be seasonally dependent. In other words, there were no meteorology × season or particulates × season interaction terms. Styer et al. found significant effects of both types.

For the remainder of this section, some of the main points of the Styer *et al.* analysis will be discussed, along with some reanalysis of the Chicago data designed to bring this more into line with the Birmingham analysis already presented.

5.1 Seasonal Parametric and Semiparametric Models

In applying Poisson regression analysis, Styer *et al.* faced the same problems of defining regressors as have already been outlined for Birmingham in Section 3. Specifically, they had to decide how to deal with trend and seasonality, which meteorological variables to include, and which measure of exposure to PM_{10}.

For the trend and seasonal variation, they used an indicator variable for each year, polynomials of up to third order for the day effect, and, in some cases, indicator variables for individual months. The year and day effects were defined separately for each of the four seasons ("spring"=March, April, May, and so on).

For the meteorological terms, the main variables used were daily mean temperature, daily mean specific humidity, and daily mean station pressure, together with their one- and two-day lagged values. For particulates, after some consideration of alternative measures, they used the three-day average of PM_{10} *including* current day, or $pmmean_0$ in our earlier notation. Note that this is different from Birmingham where the only significant effect was found to be that based on the three-day average *excluding* current day, or $pmmean_1$.

To examine interactions with season, Styer *et al.* performed two analyses, a "full-year" analysis in which the meteorological coefficients were assumed independent of season, and a "seasonal" analysis in which an entirely separate model was fitted to each season's data. They then estimated a PM_{10} coefficient for each season, and also an overall PM_{10} coefficient in the case of the "full-year" analysis.

Supplementing this was the semiparametric analysis carried out both seasonally and monthly. The main aim of this analysis was to help guide variable selection, but it was also used to estimate nonlinear effects. The separate month-by-month analyses allowed for more detailed discussion of how the various effects influencing mortality change over the year.

5.2 Results: Chicago

Data from Chicago were available for six years (1985–1990) and showed an average of 83 nonaccidental deaths per day among the elderly population. Note that this is much more data than for Birmingham, where there were 15 elderly deaths per day for 3.4 years. This increased data size is reflected in increased statistical precision of the results.

As an example of the results, Table 6 reports the seasonal PM_{10} coefficient and its standard error under the full-year regression model (a) and the seasonal

	(a) Estimate	(a) S.E.	(b) Estimate	(b) S.E.
Winter	−.00001	.00047	.00024	.00046
Spring	.00083	.00034	.00088	.00030
Summer	−.00028	.00036	−.00024	.00035
Fall	.00195	.00047	.00138	.00040
Combined	.00054	.00020		

TABLE 6. PM_{10} coefficients by season. (a) in full-year Poisson regression. (b) in seasonal model.

regression model (b). Also shown is the overall PM_{10} effect for the full-year model, which with a t value of 2.7 is statistically significant. As far as the seasonal coefficients are concerned, there is a strong PM_{10} effect in the fall and a lesser but still significant one in the spring. The summer coefficient is negative, but neither it nor the winter value are significant. Styer et al. reported that for (a), a deviance test for equality of the four PM_{10} coefficients resulted in a P-value of .001, in other words, overwhelming evidence against the hypothesis of a constant effect. They did not report a similar P-value for model (b), but rough calculations based on the estimates and standard errors in Table 6 suggest that the P-value in that case was about .01 — still significant, but not as conclusively so as for (a). The evidence of a seasonal effect complicates any interpretation of the results as indicating a simple causal relationship between particles and mortality.

Monthly analyses showed a wide range of variables appearing in the equation for different months of the year, with PM_{10} significant only in March, May, and September. Some investigation was made of the theory that these seasonal variations might be due to pollen, but detailed analysis did not support that conclusion.

5.3 Results: Salt Lake County

In Salt Lake County, data were available for 5.5 years with an average of 6.8 elderly nonaccidental deaths per day. The methodology was the same as in Chicago, with full-year and seasonal analyses and investigation of the PM_{10} coefficient for each season, as well as overall. However, none of the seasonal or overall coefficients showed a significant effect, and the overall coefficient was negative (−.00025 with a standard error of .00043). Monthly analyses showed a significant effect only for June and July, but the resulting nonlinear effect was oscillatory, the linear analysis for "summer" not being significant, just like the other seasons. From this, Styer et al. concluded that the apparent effect for June and July was probably spurious.

It is possible that the lack of statistical significance in this case simply reflects the much smaller population size compared with Chicago. However, in other studies, statistical significance has often been claimed when the population size is similar to or smaller than that of Salt Lake County. A case in point is nearby Utah Valley (Provo), in which Pope *et al.* (1992) claimed a significant effect even though the population there is half that of Salt Lake County. One plausible explanation for this discrepancy is that in Utah Valley, there is a particular known source of pollution (a steel mill), whereas in Salt Lake County, there is no such single identified source and it is likely that much of the particulate matter is natural dust. This fact illustrates the difficulties of taking results from a particular site where there is a particular source of pollution, and extrapolating them to countrywide estimates of deaths without taking into account the different forms of particulate matter.

5.4 Direct Comparisons Between Chicago and Birmingham

This section presents preliminary results from a reanalysis of the Chicago data using the same methods and models as were applied to Birmingham in Sections 3 and 4.

Weekly deaths totals are plotted in Figure 8, together with fitted curves using both the *loess* (solid curve) and the B-spline fit with one knot per month (dashed curve). The seasonal variation is significantly different from one year to another and looks more irregular than for Birmingham, which is why a higher density of knots was used in the B-spline representation.

The detailed statistical analysis is restricted to the period April 15, 1986 – December 31, 1990. The reason for this is that the observational PM_{10} record is rather sparse up to April 1986, but from then on, it is essentially a daily record with comparatively few missing days. This seems more satisfactory for an analysis which is trying to take into account, among other things, the comparative effect of different days' exposure levels.

Styer *et al.* used daily mean temperature, specific humidity, and pressure in their analysis. Here, it was decided to restrict the variables to daily mean temperature (mntp) and daily mean specific humidity (mnsh), plus lagged values of up to 4 days. Both linear and quadratic terms were considered.

In Figure 9, a plot of daily deaths against daily mean temperatures (lagged 1 day) is shown together with a *loess*-fitted trend. The plot is similar in appearance to Figure 3 except that the apparent change in slope occurs at around 22^oC, which is a different level from the 30^oC used in Birmingham. The difference could, however, be due entirely to the fact that the present plot is based on daily mean temperature, whereas Figure 3 used daily maxima. This motivates the introduction of an additional meteorological variable

$$tg22 = (mntp - 22)_+$$

together with its lagged values.

After a standard variable selection, the following variables were judged significant: trend as modeled by a B-spline representation with 56 basis functions

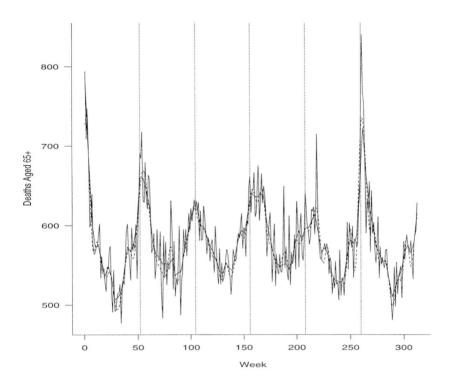

FIGURE 8. Weekly death totals for Chicago, 1985–1990, with fitted smooth curves by *loess* (solid curve) and B-spline fit (dashed curve).

(one for each month): mntp_0, mntp_1, mntp_3, mntp_4, tg22_0, tg22_0^2, tg22_1, tg22_1^2, mnsh_0, mnsh_1, mnsh_3 and mnsh_3^2. Once again, decisions were made such as not dropping a mntp variable if the corresponding tg22 was still in the model. To these variables, pmmean_0 was added. The main analysis was based on a normal linear regression fitted to log deaths. As judged by log likelihood, this actually fit better than Poisson regression. Table 7 shows the parameter estimates and standard errors. It can be seen that the pmmean_0 term is significant with a t statistic of 3.16 (P-value about .0015).

For this analysis, the comparative effects of individual days' PM_{10} have been considered in similar fashion to that of Birmingham, but by far the strongest influence was from the current day's value (lag 0). An analysis based on pmmean_1, in other words using the only form of PM_{10} variable that was significant for Birmingham, here gives a result which is not statistically significant (coefficient .00039 with a standard error of .00023; t statistic 1.71, P-value .088).

In this case, the P-value for pmmean_0 is small enough to withstand a Bon-

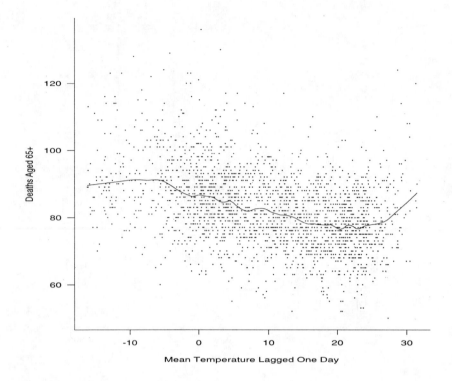

FIGURE 9. Plot of daily deaths vs. mean temperature lagged one day, with fitted *loess* smoothed curve.

ferroni multiplication factor; e.g., if we were comparing 10 different measures of PM_{10} and selected the one which gave the most significant effect, then a conservative estimate of the true resulting P-value is obtained by multiplying the P-value by 10, in this case resulting in .015, still significant. However, the inconsistency between the results for $pmmean_0$ and $pmmean_1$ is still puzzling — why should the current day's PM_{10} have such a strong influence?

Preliminary analysis of a nonlinear effect based on $pmmean_0$ suggests that in contrast to Birmingham, the effect does persist down to the lowest levels of PM_{10} and there is no evidence of a threshold other than 0. The results of this will be reported elsewhere.

A seasonal analysis may be performed by retaining the same basic model, including the B-spline representation of the trend and the same meteorological and PM_{10} terms, but adding meteorology × season and PM_{10} × season interaction terms. In the case of meteorology, this means adding $36=12\times 3$ terms to the model (12 meteorological variables multiplied by 3 df for season). The results were as follows:

Variable	Coefficient	S.E.	t Value
$mntp_0$	.00070	.00112	0.63
$mntp_1$	−.00008	.00117	−0.07
$mntp_3$	.00219	.00119	1.84
$mntp_4$	−.00211	.00084	−2.52
$tg22_0$	.01803	.00700	2.58
$tg22_0^2$	−.00228	.00096	−2.38
$tg22_1$	−.01361	.00714	−1.91
$tg22_1^2$	.00328	.00096	3.42
$mnsh_0$	.00653	.00235	2.78
$mnsh_1$	−.00532	.00234	−2.27
$mnsh_3$	−.01002	.00547	−1.83
$mnsh_3^2$	.00051	.00023	2.26
$pmmean_0$	.00073	.00023	3.16

TABLE 7. Individual coefficients and standard errors, for best linear model fitted to logarithm of Chicago elderly deaths.

(a) There is no evidence of any seasonal interaction in the meteorological variables. The reduction in deviance when the 36 interaction terms were added was 27.3, which corresponds to a P-value of .85 based on the χ^2_{36} reference distribution.

(b) The pattern of PM_{10} coefficients is similar to that found by Styer et al. (.00054, .00087, .00042, and .00137 for winter, spring, summer, and fall) but the variability is less pronounced and not statistically significant (deviance of 6.25, 3 df, P-value .100).

The result (a) is different from Styer et al., but the reason may well be the inclusion of the tg22 variable. If one tries to model the dependence of deaths on temperature in a linear model, then one would expect the slope to be positive at high temperatures in summer, and negative at low temperatures in winter. This could easily be interpreted as a seasonal interaction, whereas the model with tg22 allows for such effects in a different way.

The contrast of results over (b) could also be an effect of using different meteorological variables, but this is not clear. The one conclusion that does seem to emerge from these calculations is that the evidence for a seasonal × PM_{10} interaction, like that for PM_{10} itself, is highly model dependent.

Finally, to complete the circle of comparisons between Birmingham and Chicago, the seasonal analysis just described was also applied to the Birmingham data, starting from the models of Section 4, using $pmmean_1$ as the particulates-based variable. In this case, no evidence of seasonal interaction was found in either the meteorological or the particulates effect.

6 Conclusions: Accidental Association or Causal Connection?

The broad theme of the chapter has been to determine whether there is a significant relationship between particles and mortality after allowing for other factors, in particular long-term trends and seasonality, and meteorology, with particular reference to data sets from Birmingham and more briefly from Chicago and Salt Lake County.

We have outlined a broadly based methodology which allows for the trend to be modeled via B-splines and which also allows other variables to be included either linearly or nonlinearly. Model fitting is either by ordinary least squares applied to the square root or the logarithm of daily death count, or by maximum likelihood Poisson regression. Although serial correlation and overdispersion have been implicated in other analyses of this nature, they do not seem to be relevant factors in these two cities. Other methodologies, in particular GAM modeling and the semiparametric method of Sacks *et al.* (1989), have been considered more briefly.

In Birmingham, a statistically significant effect was found due to the three-day PM_{10} average lagged one day, but not when other measures of exposure were used. Examination of the contribution of individual days led to the conclusion of a negative effect for the current day (lag 0), positive effects for lags one and three days, and no effect for lags 2 and 4 days. The precise interpretation of these results is unclear but the overall conclusion of a significant relationship between particulates and mortality is hard to sustain. Analysis of nonlinear relationships suggested that such effects as could be detected were confined to the upper part of the range of PM_{10} values.

New analysis for Chicago showed a significant dependence on three-day average PM_{10}, but in this case, only if the average was defined so as to include the current day's reading. In this case, the effect seems strong enough that it cannot be explained away as purely an artifact of model selection, but the inconsistency of results for different PM_{10} measures is still hard to reconcile with a causal interpretation. Looking for seasonal interactions produced a pattern in the PM_{10} coefficients which is similar to that earlier found by Styer *et al.* (1995); in other words, a large effect in the fall, a smaller but still significant effect in the spring, and results which are not statistically significant for the winter and summer. However, in contrast to Styer *et al.*, the present analysis does not find that the overall difference among the four seasonal PM_{10} coefficients is statistically significant. This is interpreted as yet further evidence of the extreme sensitivity of all these conclusions to the specific model selection adopted.

The overall picture remains thoroughly confused. Other authors have suggested that the appearance of a consistent PM_{10} effect, over many cities with contrasting environments, is sufficient reason to infer a causal relationship. In contrast to that, we find sharp contrasts between different models fitted to the same set of data. Nevertheless, we do find some positive effects in both Birmingham and Chicago, and in the latter case, it is too strong to be explained away entirely as an artifact of model selection. There is still substantial reason for concern over the human health effects of particulate matter, but the widely reported claims that such effects are conclusively proved by statistical analysis do not appear to stand up to close scrutiny.

References

Chambers, J.M. and Hastie, T.J. (eds.) (1993). *Statistical Models in S*. Chapman & Hall, London.

Clyde, M. and DeSimone-Sasinowska, H. (1997). Accounting for model uncertainty in Poisson regression models: Does particulate matter really matter? Working Paper 97-06. Institute of Statistics and Decision Sciences, Duke University, Durham, NC.

Davis, J.M., Sacks, J., Saltzman, N., Smith, R.L. and Styer, P. (1996). Airborne particulate matter and daily mortality in Birmingham, Alabama. Technical Report 55. National Institute of Statistical Sciences, Research Triangle Park, NC.

Dockery, D.W., Pope, C.A., Xu, X., Spengler, J.D., Ware, J.H., Fay, M.E., Ferris, B.G. and Speizer, F.E. (1993). An association between air pollution and mortality in six U.S. cities. *New England Journal of Medicine* **329**, 1753–1759.

Green, P.J. and Silverman, B.J. (1994). *Nonparametric Regression and Generalized Linear Models: A roughness penalty approach*. Chapman & Hall, London.

Hastie, T.J. and Tibshirani, R.J. (1990). *Generalized Additive Models*. Chapman & Hall, London.

Moolgavkar, S.H., Luebeck, E.G., Hall, T.A. and Anderson, E.L. (1995). Air pollution and daily mortality in Philadelphia. *Epidemiology* **6**, 476–484.

Pope, C.A., Schwartz, J. and Ranson, M. (1992). Daily mortality and PM10 pollution in Utah Valley. *Archives of Environmental Health* **42**, 211–217.

Pope, C.A., Thun, M.J., Namboodiri, M.M., Dockery, D.W., Evans, J.S., Speizer, F.E. and Heath, C.W. (1995). Particulate air pollution as a predictor of mortality in a prospective study of U.S. adults. *American Journal of Respiratory and Critical Care Medicine* **151**, 669–674.

Roth, H.D. and Li, Y. (1996). Analysis of the association between air pollutants with mortality and hospital admissions in Birmingham, Alabama, 1986–1990. Technical Report. Roth Associates, Rockville, MD.

Sacks, J., Welch, W.J., Mitchell, T.J. and Wynn, H.P. (1989). Design and analysis of computer experiments. *Statistical Science* **4**, 409–435.

Samet, J.M., Zeger, S.L. and Berhane, K. (1995). The association of mortality and particulate air pollution. *Particulate Air Pollution and Daily Mortality: Replication and Validation of Selected Studies. The Phase I Report of the Particle Epidemiology Evaluation Project.* Health Effects Institute, Cambridge, MA, pp. 1–104.

Samet, J.M., Zeger, S.L., Kelsall, J.E., Xu, J. and Kalkstein, L.S. (1997). Air pollution, weather and mortality in Philadelphia, 1973-1988. *Particulate Air Pollution and Daily Mortality: Analyses of the Effects of Weather and Multiple Air Pollutants. The Phase IB Report of the Particle Epidemiology Evaluation Project.* Health Effects Institute, Cambridge, MA, pp. 1–29.

Schwartz, J. (1993). Air pollution and daily mortality in Birmingham, Alabama. *American Journal of Epidemiology* **137**, 1136–1147.

Schwartz, J. and Dockery, D.W. (1992a). Increased mortality in Philadelphia associated with daily air pollution concentrations. *American Review of Respiratory Disease* **145**, 600–604.

Schwartz, J. and Dockery, D.W. (1992b), Particulate air pollution and daily mortality in Steubenville, Ohio. *American Journal of Epidemiology* **135**, 12–19.

Schwartz, J. and Marcus, A. (1990), Mortality and air pollution in London: A time series analysis. *American Journal of Epidemiology* **131**, 185–194.

Smith, R.L., Davis, J.M., Sacks, J., Speckman, P. and Styer, P. (1998), Air pollution and daily mortality in Birmingham, Alabama: A Reappraisal. Technical Report (in preparation). National Institute of Statistical Sciences, Research Triangle Park, NC.

Styer, P., McMillan, N., Gao, F., Davis, J. and Sacks, J. (1995). The effect of outdoor airborne particulate matter on daily death counts. *Environmental Health Perspectives* **103**, 490–497.

Categorical Exposure-Response Regression Analysis of Toxicology Experiments

Minge Xie
National Institute of Statistical Sciences

Douglas Simpson
University of Illinois

1 Introduction

In the mid-1980s, an accident at the Union Carbide pesticides plant in Bhopal, India released the toxic gas methylisocyanate (MIC) in that densely populated region, killing more than 4000 people and injuring 500,000 others. Even today, many people in Bhopal are affected by illnesses related to that earlier exposure. This notorious industrial disaster not only forced scientists to pay greater attention to identifying and handling of hazardous chemicals but also prompted greater awareness of those common industrial products that contain hazard pollutants.

In order "to provide an ample margin of safety to protect public health," the Clean Air Act Amendments of 1990 (U.S. Congress 1990) require the U.S. Environmental Protection Agency (EPA) to develop emission standards for 189 hazardous air pollutants (HAPS). Within this context, an important consideration is quantitative risk assessment, i.e., the evaluation of quantitative relationships between exposure and response for different HAPS. Statistical considerations are important when conducting quantitative risk analyses, since many levels of uncertainty exist in the assessment process. For example, the true form of dose response to the environmental toxin is often unknown or uncertain, particularly as regards the nature of the response at low doses or how it varies across different species. Issues of low-dose and/or across-species extrapolation are of ongoing concern; development of new, modern methods for reducing uncertainty in the risk assessment process is an important concern in environmental risk analysis.

The National Institute of Statistical Sciences (NISS) became involved in a U.S. EPA quantitative risk assessment project, where available data from 40 heavily studied chemicals were used to perform quantitative risk assessments on the toxic potential of these chemicals, and to provide suggestions for future studies. In this chapter, we discuss dose-response models, and issues on risk assessment of tetrachloroethylene, one of the 40 heavily studied chemicals. Tetrachloroethylene is also known as perchloroethylene or PERC. It is used heavily in dry cleaning as a solvent, and in silicone lubricants, aerosol cleaners, fabric finishers, and many other applications. It is known that exposure to tetrachloroethylene can cause varied responses, including central nervous system effects such as dizziness, headache, nausea, and, in extreme cases, unconsciousness and death. Evidence from laboratory animal experiments also suggests that chronic exposures to tetrachloroethylene may induce mammalian carcinogenesis.

1.1 Critical Exposure-Response Information and Modeling Approaches

Two important exposure measurements are critical in characterizing toxic responses: exposure concentration and exposure duration. These depend on the specific source and scenario of the exposure. For example, it is well known that smaller doses over long, chronic exposure periods can have substantial toxic effects. In quantitative risk assessment, the analysis of risks should include information on exposure concentration as well as duration. Currently, the analysis of risks conducted according to the U.S. EPA relies primarily on the no-observed-adverse-effect level (NOAEL) approach (U.S. EPA 1991) and the benchmark dose (BMD) approach (Crump 1984; 1995; U.S. EPA 1995). To date, the NOAEL and BMD approaches have been used primarily to determine a dose-response relationship in a single study with fixed exposure durations. Theoretically, the BMD approach can model both exposure concentration and duration, although doing so requires a well-designed, large experiment. Even though many studies have been conducted on potentially hazard substances over the past 50 years, in practice, general methodologies and databases are missing for the study of exposure concentration and duration effects simultaneously, since available toxicological data providing detailed exposure-response information is limited. For example, recently the U.S. EPA compiled a database of acute toxicity data (exposure < 24 hours) for 40 heavily studied chemicals from the open literature (Guth and Raymond 1996). According to Guth *et al.* (1997), of a total of 890 experiments recorded (489 references), only 45% had more than 2 concentrations with quantitatively reported endpoints (half were studies of lethality). Some of these studies would be useful for assessing potentially toxic effects, and thus useful for BMD analysis. However, only six of the experiments varied concentration and duration and five of these were lethality studies.

Individual studies of the sort performed by the EPA (above) often lack of substantive exposure/duration information. Modern statistical methods are required that can combine the exposure information from individual studies. Indeed, collecting together overall exposure-response information from different laboratory studies is a critical component for public policy officials who manage the population risk of exposures to toxic chemicals. Combined results from multiple studies summarize overall associations, and inferences from the combined results are more robust than inferences from any single study. In addition, the combination of exposure-response data from different species provides quantitative connections between species for cross-species extrapolations.

1.2 Issues in Exposure-Response Risk Assessment

We discuss in this chapter a general statistical approach that combines the information from disparate studies. The focus of our discussion is on application to risk quantification with noncancer outcomes, with a case study of tetrachloroethylene serving as an example. It will be seen that the data are, in fact, quite complex, and proper data analysis for their combination is not a straightforward effort. The development of this approach has involved extensive discussions with U.S. EPA toxicologist Dr. Daniel Guth. Some of the results have been published separately; see Simpson *et al.* (1996a, 1996b), Guth *et al.* (1997), Xie *et al.* (1997), and Xie and Simpson (1998).

Some questions of interest are the following:

- Combine measurements from diverse studies on different biological outcomes and toxic endpoints;

- Model severity of toxic response when the data are Interval-censored;

- Incorporate systematic differences, such as species and gender differences, into the statistical analysis, and construct appropriate models for cross-species extrapolation;

- Adjust for nonsystematic variations that are not identifiable, such as study-to-study or group-to-group variations;

- Model responses reported at group levels.

In Section 2, we describe the tetrachloroethylene database and discuss our effort on combining diverse measurements from this information. We center attention on *severity scoring* and *interval censoring*. In Section 3, we present statistical models that are applied to the combined tetrachloroethylene data set and address different statistical issues that arise when combining the study information. We include a regression application of *stratification* for eliminating bias and improving the precision of statistical inferences, and describe a so-called *marginal modeling* technique for dealing with correlated observations and overdispersion. In Section 4, we introduce a special software program, *CatReg*, that provides both standard and nonstandard statistical computations required in our analysis. In Section 5, we summarize an application on tetrachloroethylene reported previously in Guth *et al.* (1997) and Simpson *et al.* (1996a). Section 6 discusses conclusions, and Section 7 suggests some areas for future work.

The reader is referred to the *web companion* for specific data sets and software that are related to the case studies in this chapter.

2 The Tetrachloroethylene Database

The database employed here contains information from 20 studies on central nervous system (CNS) effects after acute exposure to tetrachloroethylene. Here, "acute" refers to single exposures of less than 12 hours. As might be expected, the resulting collection represents a variety of complex toxicological endpoints and forms of response.

Initially, a literature search found all available data from published sources, proceedings, and technical reports, and these were screened carefully to remove any poorly documented studies. After omitting one report with limited information on dosages, and removing several studies with no information on CNS effects, 12 references containing 20 studies were left. The critical information recorded in the toxicological study databases is summarized below. (See also, Guth et al., 1997.)

Study/group index References, experiment (study), group number

Exposure information Concentration, duration, postexposure duration

Original response Target organ (CNS), effect description, type of data

Covariates Species, strain, sex

Severity category Designation as 0, 1, or 2 (NE, AE, or SE)

Censoring Range of severity, best judgment for censoring information

Response detail index (Response reported is at) group or individual level

Each of the 20 studies contained a certain number of experimental groups with similar exposure concentrations and durations. There were total 118 experimental groups, each of which had a certain number of experimental subjects. Some of studies reported information only at group level with the numbers of experimental subjects in the groups specified; others had information at the individual level. The experimental subjects included mice, rats, and human beings. Under the gender category, we define *male*, *female* and *not specified*, since a fair amount of data did not specify the sex of the subjects. Table 1 lists profile information for the tetrachloroethylene database.

2.1 Severity Scoring

The original response measurements from the literature were quite different; for example, a considerable number of human studies reported subjective symptoms, while some of the animal experiments reported numerical responses such as defined stages of anesthesia or percentage changes in time an animal was immobile. Based on previous works by Hertzberg and Miller (1985), Hertzberg (1989), and Guth et al. (1991), for each study the responses on groups or

	Number of Groups		
Species	Male	Female	Unspecified
Mouse	17	35	12
Rat	5	7	22
Human	14	0	6

TABLE 1. Summary of toxicological data available for tetrachloroethylene Unspecified includes experiments with sex not specified and with combined male and female subjects. [A modified version of Table 2 from Guth et al. (1997).]

individual subjects (if possible) were reduced to an ordinal scale of severity categories: No-Effect (NE), Adverse-Effect (AE) and Severe-Effects (SE). A NE is an effect that is either not different compared with the control or is different but judged not to be adverse; an AE is an effect that is judged to be adverse; and a SE is an effect of lethality or judged to be nearly lethal (Guth et al. 1997). These scores of severity are similar and related to severity categories developed by the NRC (1993), Rusch (1993), and Cal EPA (1994). We call this process *severity scoring*. A distinct advantage of adverse outcome modeling through severity scoring is that it provides a way to place very different quantitative measurements on a common scale (Simpson et al., 1997).

The severity scoring was performed by a toxicologist and the scores were recorded in the database. In addition, various endpoints associated with scores were interpreted and prepared for inclusion in the database by technical experts. The scoring was based on biological considerations rather than on statistical significance, since it was felt that in environmental toxicology, statistically significant changes may be biologically unimportant and biologically important effects may not be statistically significant. Of course, in practice, statistical significance plays a major role in determining adversity, especially when there is limited basis in the biological theory for a judgment. This practice introduces an important bias into the analysis, because the same level of response could be considered adverse or not adverse, depending on the number of animals or subjects in the group (Guth et al., 1997). Since a limited number of studies is considered in our analysis, this source of bias could have a large effect if we did not enforce the biological consideration.

2.2 Censoring

Because there is often insufficient information to determine the biological significance in a particular response, substantial censoring occurs in severity scoring. For example, if the level of toxicity associated with a continuous response (such as percentage change in duration immobility) is in fact not known or not agreed upon, severity information may have to be censored; i.e., the response is assigned to two or more adjacent severity categories.

Another type of censored observation comes from incomplete information.

We have a few lethality studies in the tetrachloroethylene database which target fatal exposures and have outcomes reported merely as survival or death. This information on survival does not provide further scoring on NE or AE. Such incomplete information in severity category is approached by treating the effect as censored.

3 Statistical Models for Exposure-Response Relationships

Statistical analysis and regression modeling are central to quantitative risk analysis. Often, responses are regressed on exposure concentration and exposure duration. The types of responses could be either continuous (e.g., weight losses) or categorized scores (e.g., severity scores). Models used to fit categorical responses are called *categorical regression models*. In noncancer risk assessment, various levels of effect severities are seen. For instance, the NRC (1993), Rusch (1993), and Cal EPA (1994) discussed different effect severities corresponding to mild or transient effects, adverse or disabling effects, and severe or life-threatening effects. In similar fashion, in our tetrachloroethylene database we consider three ordinal levels for scoring severity: no effect, adverse effect, and severe effect. To analyze these categorized severities, categorical regression analysis is appropriate.

In Section 3.1, we first introduce an empirical exposure-response law know as *Haber's law* and use it as a justification for employing (categorical) linear regressions. We then discuss a series of statistical models stemming from Haber's law. In Section 3.2, we discuss the class of homogeneous categorical regression models and their application to interval-censored observations, and formalize a definition of *effective dose*. In Section 3.3, we introduce a *stratification* regression technique, which is used to model any systematic difference such as gender or species differences. In Section 3.4, we sketch the so-called *marginal modeling* technique, which may be used to deal with non systematic study-to-study and group-to-group variation. Finally, in Section 3.5, we briefly address some additional issues that arise in the analysis of the tetrachloroethylene data. These issues include group-level reporting, nonzero control responses, and sensitivity of the severity scoring.

3.1 Haber's Law

An empirical law know as *Haber's law* and its modifications roughly describe the toxicological interrelationship between exposure concentration and duration; see, e.g., Atherly (1985). If we denote exposure concentration as C and exposure duration as T, then Haber's law is expressed as $C^\gamma \times T =$ CONSTANT TOXICITY. It states that equal products of $C^\gamma \times T$ cause the same toxicity. The quantity γ is a hyperbolic parameter that determines the shape of a (equal toxicity) response curve, and it varies from chemical to chemical.

Empirical studies from ten Berger et al. (1986) suggest that for lethality data, γ ranges from 0.8 to 3.5 for acute exposures of 90 minutes or less. Haber's law is based essentially on characterization of the exposure-response continuum at a simple level. It does not carefully account for mechanistic determinants of the disposition of a parent compound and its metabolites (Jarabek 1995). Although more complicated dosimetry models, such as physiologically based pharmacokinetic (PBPK) models (Ramsey and Anderson 1984), have been proposed to provide more detailed connections between dose and toxicity, the use of Haber's law is considered acceptable "in the absence of sufficient mechanistic data" (Jarabek 1995).

If we take logarithms, we have a modified Haber's law: $\gamma \log(C) + \log(T) =$ CONSTANT LOG-TOXICITY. Notice this is now linear in γ. For the data studied herein, Haber's law serves as a modest justification for us to use a linear regression model with $\log(C)$ and $\log(T)$ as explanatory variables. We might be able to assess the adequacy of Haber's law through statistical testing of the linearity of the regression covariates; however, we have not carried out this effort in our analysis to date.

3.2 Homogeneous Logistic Model

Assume that three ordered severity categories are labeled as $0, 1$, and 2, corresponding to No-Effect (NE), Adverse-Effect (AE), and Severe-Effect (SE) in the tetrachloroethylene data. Let Y be the ordinal severity category for a particular observation, and let $x_1 = \log(C)$ and $x_2 = \log(T)$ be the observation's exposure history. A common model for ordinal data is polytomous logistic regression, where the ordered outcomes are viewed as the polytomous categories, and a logistic function is used to describe the probability of response as a function of x_1 and x_2. To model the severity score Y in this manner, we consider

$$\Pr(Y \geq s | x_1, x_2) = \begin{cases} H(\alpha_s + \beta_1 x_1 + \beta_2 x_2), & \text{if } s = 1, 2 \\ 1, & \text{if } s = 0 \end{cases} \quad (7.1)$$

where $H(t) = e^t/(1+e^t)$ and $H^{-1}(p) = \log\{p/(1-p)\}$ is the logit of p. Model (7.1) entails a proportional odds ratio assumption for the different severity categories, and is known as a *proportional odds ratio model*; see McCullagh (1980) or Agresti (1984). In the literature of generalized linear regression models (McCullagh and Nelder 1989), $H^{-1}(p)$ is called the *link function* and the form in (7.1) is a *logit model*. Simpson et al. (1996b) discuss other possible choice of regression models for ordinal observations. Xie (1996) describes a latent structure for model (7.1) in the dose-response context.

As mentioned in Section 2.2, in the tetrachloroethylene data we sometimes do not have enough information to judge if a response Y is a no-effect or an adverse effect, and in this case, we view Y as (interval) censored. If the responses Y are (interval) censored and their upper and lower limits are known, say a and b where $a \leq b$ take values $0, 1, 2$, we can through model (7.1), calculate the probability of observing the censored response as

$$\begin{aligned}\Pr(a \leq Y \leq b|x_1, x_2) &= \Pr(Y \geq a|x_1, x_2) - \Pr(Y \geq b+1|x_1, x_2) \\ &= H(\alpha_a + \beta_1 x_1 + \beta_2 x_2) - H(\alpha_{(b+1)} + \beta_1 x_1 + \beta_2 x_2).\end{aligned}$$

Note that if $a = b$, then this probability reverts to model (7.1). Let us denote the likelihood function associated with this model as $l(\mu|\mathbf{Y})$, where $\mathbf{Y}$ is the vector of observations and $\mu = \mu(\alpha, \beta) = E(\mathbf{Y})$. If the observations are independent, we compute $l(\mu|\mathbf{Y})$ by multiplying together the probabilities of each observed response, that is,

$$l(\mu|\mathbf{Y}) = \prod_v \{H(\beta_1 x_{1v} + \beta_2 x_{2v} + \alpha_{a_v}) - H(\beta_1 x_{1v} + \beta_2 x_{2v} + \alpha_{b_v+1}),\}$$

where the index v indicates each single observation. By maximizing the likelihood, we obtain maximum likelihood estimates for α's and β's; Simpson et al. (1996b) give an extended discussion on censored models.

One important concept often seen in categorical regression analysis is the use of (*scaled*) *model deviance*. The (scaled) deviance is defined as twice the difference between the maximum log likelihood achievable under a fully saturated model, $l(\mathbf{Y}|\mathbf{Y})$ (i.e., replacing μ with $\mathbf{Y}$ in the likelihood function), and that achieved by the model under investigation, $l(\mu|\mathbf{Y})$. The deviance is commonly written as $-2\{l(\mu|\mathbf{Y}) - l(\mathbf{Y}|\mathbf{Y})\}$ (McCullagh and Nelder, 1989). In a standard normal linear regression model for continuous observations, the model deviance corresponds to the common residual sum of squares (divided by its error variance σ^2). In both continuous and categorical models, the corresponding deviances provide a generalized likelihood ratio type test for the fitted models, and the computations on fitting model (7.1) can be carried out through standard statistical software packages such as SAS^{TM} (SAS Institute Inc.), SPSS^{TM} (SPSS Inc.), S-PLUS^{TM} (MathSoft Inc.), etc. The output from these standard packages includes parameter estimates, associated variance–covariance matrix estimates, and model deviances, etc., where the variance–covariance matrix estimate is often the inverse sample Fisher information matrix (the second derivative of the log likelihood function with respect to the parameters). Unfortunately, the standard packages do not support this type of computation for interval censored data. Special software to carry out the censored-data calculation is discussed in Section 4 below.

Model (7.1) yields a formula for the $100q\%$ *effective dose*, ED_{100q}, for severity categories $s = 1, 2$, i.e., the hypothetical dose level which would lead to 100% rate of response as severe as category s. On a logarithmic scale, the effective dose for fixed duration x_2 is

$$\text{ED}_{100q}(x_2) = \frac{\text{logit}(q) - \alpha_s - \beta_2 x_2}{\beta_1}.$$

For a given duration, lower values of ED_{100q} correspond to more potently toxic chemicals. If the chemical is toxic, a plot of ED_{100q} versus the log-duration yields a downward sloping line; hence smaller doses can induce more potent toxicity if the duration of exposure is increased. Inserting estimates for the unknown parameters yields an estimate for the ED_{100q} line. (See, Carroll et al., 1994)

To understand the ED_{100q} line, consider the adverse ED_{100q} line for rats. Suppose there are 100 rats exposed to tetrachloroethylene at a certain set of exposure dosage and duration on the ED_{100q} line. Of these 100 rats, we expect on average that $100q$ will exhibit adverse or severe effects.

Often, the delta method (Cox and Hinkley 1981) is used to find approximate pointwise confidence intervals for ED_{100q}. More specifically, denote the parameter estimates of $(\alpha_s, \beta_1, \beta_2)$ as $(\hat{\alpha}_s, \hat{\beta}_1, \hat{\beta}_2)$, their variance–covariance matrix estimate as $\mathbf{V}$, and $\mathbf{a}^T = (-1, \hat{ED}, -x_2)/\hat{\beta}_1$, where $\hat{ED}$ is the estimate of effective dose at duration x_2. Then, according to the delta method, a variance estimate for $\hat{ED}$ is $\{\mathbf{a}^T \mathbf{V}^{-1} \mathbf{a}\}^{-1}$. Two other commonly used methods to find the confidence intervals for effective doses are Fieller's fiducial limits and likelihood ratio confidence bands. Interested readers may find discussions of the first in, e.g., Finney (1971) and Morgan (1992), and the second, e.g., in Alho and Valtonen (1995).

3.3 Stratified Regression Model

The underlying assumption of model (7.1) is homogeneity; i.e., conditional on the exposure, the response scores are the result of a purely random multinomial sampling process. The assumption of a purely random multinomial error distribution across studies is difficult to defend in our setting, however, scientifically or statistically. In order to investigate the systematic sources of heterogeneity, such as gender and species differences, in the database, we follow Simpson et al. (1996b), who allowed certain parameters to vary between different subsets of the data. For instance, we allow different experimental species to have their own parameters. These subsets are called *strata*, and they are constructed with the expectation that the data they contain are more homogeneous than the database as a whole. In the design and analysis of experiments, stratification is an important and powerful concept for eliminating bias, decreasing variation, and, hence, improving both the accuracy and precision of statistical inferences.

A stratified regression model generalizes the scope of the problem by allowing for subgroup differences. It still assumes that other differences are due to multinomial random sampling, however, so we expand model (3.2) to include distinct regression parameters $(\alpha_1^{(j)}, \alpha_2^{(j)}, \beta_1^{(j)}, \beta_2^{(j)})$ for stratum j, $j = 1, \ldots, J$, corresponding to different species, sexes, etc. The expanded model is given by

$$\Pr(Y \geq s | x_1, x_2, \text{stratum } j) = \begin{cases} H(\alpha_s^{(j)} + \beta_1^{(j)} x_1 + \beta_2^{(j)} x_2), & \text{if } s = 1, 2 \\ 1, & \text{if } s = 0. \end{cases} \quad (7.2)$$

Under (7.2) and for $H(t) = e^t/(1+e^t)$, the formula for effective dose corresponding to stratum j becomes

$$\text{ED}^{(j)}_{100q}(x_2) = \frac{\text{logit}(q) - \alpha_s^{(j)} - \beta_2^{(j)} x_2}{\beta_1^{(j)}}.$$

Every stratum in model (7.2) is structured to have its own set of parameters. Of course, this is not always necessary since there may be mechanistic similarities between certain strata. In this case, some type of parameter sharing might be possible across the strata. Statistical testing of nested models provides a means to determine such parameter sharing, and therefore yields better insight into the exposure-response mechanism. Model testing and selection within this framework is commonly directed by a series of likelihood ratio (model deviances) tests between nested models.

As an example, if the strata share the same slope parameters β_1 of concentration and β_2 of duration, and the stratification is incorporated only via the intercept term, then model (7.2) reduces to

$$\Pr(Y \geq s | x_1, x_2, \text{stratum } j) = \begin{cases} H(\alpha_s^{(j)} + \beta_1 x_1 + \beta_2 x_2), & \text{if } s = 1, 2 \\ 1, & \text{if } s = 0. \end{cases} \quad (7.3)$$

Under (7.3), the (equal toxicity) response curves across different strata will be parallel to one another if plotted on a log-scale. The difference between them is that different strata have different sensitivities. As we will see in Section 5, statistical analysis of the tetrachloroethylene data indicates that rats and humans are likely to share concentration and duration parameters, and that the rat ED_{10} line is essentially parallel to the human ED_{10} line. However, the rat ED_{10}'s are 10- to 100-fold larger than the human ED_{10}'s. This suggests that rats and humans may have similar exposure-response mechanisms, but humans seem more sensitive to tetrachloroethylene than rats (although this sensitivity can be due either to biological differences between human beings and rats, or the different way of measuring the responses, or both). Understanding this type of mechanism can be helpful for across-species extrapolations.

Models (7.2) and (7.3) are essentially proportional odds ratio models with more regressors than models (7.1). They can be fit using standard software, e.g., SAS Proc Logistic, in the case of no censored observations; otherwise, the models are fit using the specific software discussed in Section 4 below. The structure of these stratified models is very general, accommodating many different kinds of partial stratification and covariate information. These include stratification of the risk on species, sex, endpoint category, etc. It should be noted, however, that multiple testing between a number of models inflates the false positive error rate, so it is possible to achieve nominally statistically significant differences that are nonsignificant after adjustment for multiplicity.

3.4 Marginal Modeling Approach

In the tetrachloroethylene database, there are also nonsystemic sources of variation. As Simpson et al. (1996b) point out, our statistical analysis is complicated by two factors: (a) data are collected typically by grouping experimental individuals together (in the tetrachloroethylene database, the median group size is 6) and (b) groups are clustered within studies. These concerns lead to correlation among the observations. As a result, one can expect to observe overdispersion in the observed counts. Overdispersion among responses within a study can bias estimated standard errors unless proper adjustment is made; see McCullagh and Nelder (1989, Sec. 4.5) for an account on overdispersion.

If one thinks of the separate studies as clusters, one sees that we are within the typical framework of a generalized estimating equations (GEE) approach (Zeger and Liang 1986); that is, we specify that an individual response has marginal distribution (7.1) or (7.2) or (7.3), given the observed covariates, and we assume responses within a cluster are correlated with an unspecified correlation structure. A *pseudo-likelihood* is constructed by treating the observations as if they are independent, that is, by multiplying together the individual marginal distributions. This pseudo-likelihood has the exact form as the likelihood expression for independent observations, such as (3.2). The estimates obtained by maximizing the pseudo-likelihood function are asymptotically consistent for the true parameters; see, e.g., Li (1996). However, standard formulas to calculate variance estimates (i.e., via the Fisher information matrix from standard likelihood inference as described in the previous subsections) are incorrect. Instead, we generalize the standard Fisher information matrix formula to a "sandwich" formula $A^{-1}BA^{-1}$, where A is the expected matrix of negative second derivatives of the pseudo-likelihood function with respect to all parameters, and B is the covariance matrix of the first derivative vector of the pseudo-likelihood function with respect to all parameters. If the observations are independent, both A and B will equal the Fisher information matrix; see Simpson et al. (1996b).

In the tetrachloroethylene data, group size and study size vary from one experiment to another. To improve estimating efficiency in the marginal approach, we allow for weighting of individuals within studies and groups, as well as for weighting of studies and groups. In our analysis, we weight the term of the pseudo-likelihood corresponding to the observation from the jth group in the ith study by $\{1/(\text{size of } i\text{th study})\} \times \{1/(\text{size of } j\text{th group})\}$. This scheme down-weights the single large groups or single large studies. Simpson et al. (1996b) illustrated that this type of down-weighting improves the efficiency of the estimator and gives effective doses narrower confidence bands compared with an unweighted approach.

Marginal models and generalized estimating equations have been the subject of a large body of research. As two classes of modeling approaches to correlated observations, the major difference of this modeling approach from the classical random effects model (see, e.g., Lehmann 1983) is that the marginal approach targets the population average distribution rather than the individual observations. In marginal modeling, there is no assumption made on correlation structures, thus it is, in a sense, more general than a random effects model. If, however, the random effects models are specified correctly, estimation and inferences from a marginal analysis are often less efficient than those from a random effects model. In the analysis of combined toxicology data, we recommend the marginal approach, since correct specification of the correlation structure is difficult for such data, and estimates in a marginal analysis are fairly robust to incorrect correlation specifications. Moreover, as a practical matter, random effects models are often much more difficult to compute. Simpson et al. (1996b) and Xie (1996) discuss the relationship between marginal modeling and classical random effects (or conditional) modeling, as applied to estimation of effective doses. They note that under a random intercept model, the effective doses defined under the marginal model are the same as the ones defined under the random effects model. For readers interested in more details on the marginal modeling approach, a systematic statistical discussion can be found in Diggle et al. (1994).

3.5 Other Issues

Toxicological data often have complex, heterogeneous structures, leading to numerous practical and statistical issues associated with the modeling approach presented above. Due to space limitations, we only summarize these issues here. Interested readers can find further details in the cited references.

Grouped Data

The use of information from the literature and other reports, rather than from original data, means that some data will be available at the group level only. As a result, some of the severity scores can only be assigned at the group level, thus there is a combination of group level information for some experimental subjects, and individual information for the remainder. The existence of group-level responses as well as individual-level response requires nonstandard modeling. Xie et al. (1997) outline a statistical modeling approach for the analysis of group-level responses when the group sizes vary. They use a latent random effect structure to model the group-level responses whenever individual information is not reported, and the within-group correlation for the groups is estimated. In their latent model, an individual-level response can be viewed as a special case of a group level response where group-size equals one. In a preliminary analysis of the tetrachloroethylene data, their results suggest that the within-group correlation is so high that a grouped response might reasonably be considered to reflect the response of a single individual.

Control Observations

Another issue is the proper handling of control (i.e., zero dose) observations. In our modeling approach, the explanatory dose variables are taken over a logarithmic scale. If the responses in the control groups are all null responses, we can simply ignore the control data, since they are uninformative under our log-dose model. But any non-null responses among the controls would, strictly speaking, invalidate the log-dose model. For binary regression, Abbott's formula provides a relatively clean analysis by adding a baseline reference ratio into the probability model; see, e.g., Morgan (1992, Section 3.1). Let λ be the baseline reference ratio and $\Pr(d)$ be the probability of overt exposure response to dose d. Then, Abbott's formula models the observed response probabilities with a nonzero baseline model $\lambda + (1-\lambda)\Pr(d)$. Noticing, in the Abbott's formula, λ corresponding to responses due to natural causes and $\Pr(d)$ due to exposure causes, Xie and Simpson (1998) extend this binary nonzero baseline model to ordinal cases. They describe a simple (and biological meaningful) latent structure for the model, where they distinguish latent responses due to natural or spontaneous causes from latent responses due to exposure to experimental stimuli. Both types of latent responses are modeled via multinomial distributions, and the observed response probability is a mixture of these two multinomial distributions. This latent structure provides an ordinal nonzero baseline model, and an EM algorithm (Dempster et al. 1977) is used to compute the unknown parameter estimates.

Worst-Case Analyses

Our database for the combined studies contains a range of toxicological endpoints, a range of subjects, and a range of experimental design points, as well as studies with a range of different experiment goals. This diversity poses a challenge to our analysis. To ensure that our inferences are relatively insensitive to small changes in the modeling procedure, we refer to Simpson et al. (1996a), who present a worst-case analysis as an alternative approach to the interval-censoring approach. In the worst-case analysis, uncertain or unknown severity cases are resolved in favor of the most severe of the known severity levels; i.e., the interval censored scores in the database are assigned worst-case values. For example, in the tetrachloroethylene data, if a response score is interval-censored among NE and AE, it is assigned to AE under the worst-case scenario. Comparing the worst-case with the censored analysis provides an indication of the scoring sensitivity of the analysis. Simpson et al. (1996a) also explored the stability of the parameter estimates in the worst-case analysis by varying a robustness parameter, and graphing the curves traced out by the estimates and confidence intervals.

4 Computing Software: *CatReg*

In the previous section, we discussed some standard and nonstandard statistical approaches needed in our risk assessment analysis. As we noted, standard software cannot provide all the computations we need. To address this concern, we constructed our own software package, *CatReg*, to carry out these computations. *CatReg* is a customized software package that runs under the S-PLUS computing environment. It provides diagnostics for model selection and model criticism, and a query-based interface for data analysis and graphics. Its model fitting output is formatted similarly to standard S-PLUS glm function output. The main features of *CatReg* are listed below:

- *CatReg* provides maximum likelihood estimates when the observations are independent, and provides consistent maximum pseudo-likelihood estimates when the observations are batch correlated, where the standard errors are adjusted by the method of generalized estimating equations (GEE).

- *CatReg* has the ability to handle a variety of common link functions for continuous, binary, or ordinal data.

- *CatReg* handles interval-censored observations, which were discussed in Section 3.2.

- As an option, *CatReg* can perform a worst-case analysis by automatically assigning the worst scores to the severity responses (as described in Section 3.5).

- *CatReg* allows simultaneous modeling of group- and individual-level responses via a scaling option.

- *CatReg* package provides computation and graphic output for effective dose, ED_{100q}, estimation.

Interested readers can find detailed information about *CatReg* and its installation instructions in documentation by Simpson *et al.* (1995), and also in the web companion to this chapter.

5 Application to Tetrachloroethylene Data

In order to study the structure of the tetrachloroethylene data in more detail, Guth *et al.* (1997) considered a series of models. These included the homogeneous model (7.1), the stratified regression model (7.2) with common concentration and duration slopes, and the stratified model (7.2) with common concentration slope. In the stratified models, the intercepts were stratified by

	Model	Deviance	parameters	P-values
1	Homogeneous	283.74	4	—
2	Common slope, intercept by species	281.21	6	$> 0.05^a$
3	Concentration slope, intercept by species	252.40	6	$< 0.001^b$
4	Common slope, intercept by species and sex	274.98	11	$> 0.01^b$
5	Concentration slope by species intercept by species and sex	241.56	18	$> 0.05^c$

TABLE 2. Comparison of a hierarchy of models applied to central nervous system effects of PERC in rats, mice and humans. * P-values are evaluated through a weighted sum of χ^2 distributions (Xie, 1996): ^{a}compared with model 1, ^{b}compared with model 2, and ^{c}compared with model 3.

species and by both species and sex. Table 2 lists model deviances for fitting a hierarchy of nested models, where we took interval censoring into account and adopted a working hypothesis that group-level response scores were equivalent to scores of single individuals. According to Table 2, stratification of intercept and concentration slope on species significantly improved the model fit (model 3 in Table 2, P-value < 0.001), while gender stratification did not provide further significant improvement over species stratification (model 4 in Table 2, P-value > 0.01). Figure 1 plots the adverse effect ED_{10} lines under model 3 with intercept and concentration slope stratified on species. It indicates that the concentration slope parameters for humans and rats appear parallel and that they are both different from the mice. Further statistical testing results (not shown here) confirm this point. We note that the confidence intervals for the mice stratum parameters, including ED_{10}, are quite large. Careful examination of the data set reveals that there is a lack of design points for mice at lower exposure levels.

A worst-case analysis resulted in slightly lower ED_{10} values (Figure 2). At higher doses, the ED_{10} values for rats, mice, and humans are close to those obtained by using the interval-censored approach (Figure 1); at lower doses, they are different by a factor close to 2. The flatness of the ED_{10} lines and their wide confidence bands at low durations in the worst-case analysis reflect the weak duration effect for the range of data included in the analysis; notice that responses known to be nonadverse occurred only over a narrow range of (high) durations under this analysis. Again, Simpson et al.'s (1996a) results indicate that the concentration slope parameters for humans and rats are parallel and that they both differ from the mice. The confidence intervals for the mice

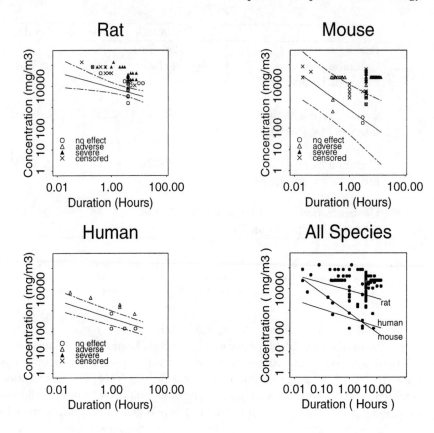

FIGURE 1. ED_{10} (solid lines) and 95% confidence bounds (dashed lines) versus duration. Interval-censored analysis (a slightly modified version of Figures 2 and 3 from Guth et al. 1997).

stratum parameters, including ED_{10}, are also quite large. A robust analysis conducted by Simpson et al. (1996a) studied the sensitivity of the model fit to uncertainty in the stratum definitions. They found that the mice stratum parameters are quite unstable, and, in particular, that poor information at low doses could exacerbate this poor behavior.

6 Conclusions

Data available on acute exposures to tetrachloroethylene provide a limited basis for quantitative risk assessment, using conventional approaches such as the no-observed-adverse-effect level (NOAEL) or the benchmark dose (BMD). For the available tetrachloroethylene studies, the NOAEL approach may be applied at several durations in some studies, and the BMD approach could be adopted for a single study with a fixed duration. However, the limited experimental designs in the available data do not allow for the study of concentration

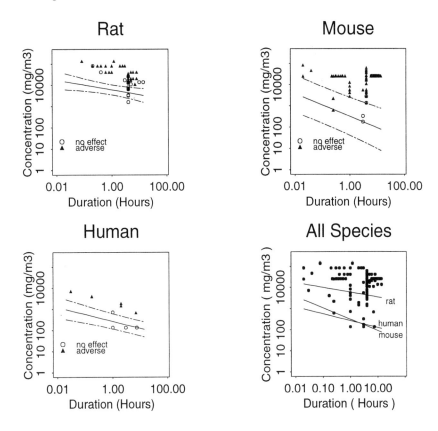

FIGURE 2. ED$_{10}$ (solid lines) and 95% confidence bounds (dashed lines) versus duration. Worst-case analysis for the tetrachloroethylene data (a slightly modified version of Figure I from Simpson et al. 1996a).

and duration simultaneously. By incorporating data from all adequate studies, our combined approach provides a useful alternative to standard exposure assessments.

Comparison of a series of nested models allows us to evaluate the important sources of variability in the database and to select the best model for use in risk assessment. If the implicit assumption that linkage between severity categories and between strata is acceptable, then estimating human effect probabilities at the severe effect categories is possible by combining species in one single analysis (Guth et al., 1997). Comparing between nested stratified models provides a method for testing significance of the difference between strata. Differences between strata may point to important differences in mechanism, dosimetry, or sensitivity; similarities between strata would support data combination and extrapolation. It is important, though, to recognize that differences between model predictions for different species may be due in part to differences in sensitivity of the reported endpoints, as well as to other species differences (Simpson *et al.*, 1997).

7 Future Directions

Data generated to model dose response and assess risk of toxic exposures to environmental chemicals involve complicated structures and interrelationships. In addition to the methods described above, more complex dosimetric approaches such as physiologically based pharmacokinetic (PBPK) models (Ramsey and Anderson 1984) are of additional research interest. PBPK models are a class of dosimetric approaches that focus on how physiology and kinetics influence the way a substance travels through the body. They involve a large amount of experimental data, and an extensive set of differential equations is required to incorporate these features. With these, scaling of mechanistic parameters such as metabolic rates provides for accurate extrapolation from animal to humans.

We can use the combining information approach presented in this chapter to enhance PBPK-based risk assessment. The idea is to use toxicological data to obtain indirect measures of the efficacy of PBPK modeling for cross-species extrapolation. Note that the PBPK model implies a method for computing human equivalent concentrations (HECs) when using animal data to predict human toxicological response. We can regress the severity response score on the HEC derived from a PBPK model. Then, using methods developed in this chapter, we test whether stratification on species and other factors is still necessary after the HEC adjustment. If the PBPK model is correct, the HEC adjustment will reduce the size of interspecies adjustments required by the pooled data analysis. Within our modeling framework, it will be possible to test the significance of species differences after HEC adjustment, and thus assess whether a PBPK justification is adequate.

Acknowledgments

This chapter is based on discussions, publications, and technical reports by Professor Raymond J. Carroll, Dr. Daniel J. Guth, Professor Douglas G. Simpson, Dr. Minge Xie, and Dr. Haibo Zhou in a NISS-EPA risk assessment project. The authors wish to thank Professor Jerome Sacks for his management role in the project and for his valuable suggestions.

References

Agresti, A. (1984). *Analysis of Ordinal Categorical Data*. Wiley, New York.

Alho, J.M. and Valtonen, E. (1995). Interval estimation of inverse dose-response. *Biometrics* **51**, 491–501.

Atherly, G. (1985). A critical review of time-weighted average as an index of exposure and dose, and of its critical elements. *American Industrial Hygiene Association Journal* **46**, 481–487.

California Environmental Protection Agency (1994). *Guidelines for Determining Acute Chemical Exposure Levels* (Public Comment Draft). Office of Environmental Health Hazard Assessment, Hazard Identification and Risk Assessment Branch, Berkeley, CA.

Carroll, R.J., Simpson, D.G. and Zhou, H. (1994). Stratified ordinal regression: A tool for combining information from disparate toxicological studies. Technical Report #26. National Institute of Statistical Sciences. Research Triangle Park, NC.

Cox, D.R. and Hinkley D.V. (1981). *Theoretical Statistics*. Academic Press, London.

Crump, K.S. (1984). A new method for determining allowable daily intakes. *Fundamental and Applied Toxicology* **4**, 854–871.

Crump, K.S. (1995). Calculation of benchmark doses from continuous data. *Risk Analysis* **15**, 79–89.

Dempster, A.P., Laird, N.M. and Rubin, D.B. (1977). Maximum likelihood from incomplete observations. *Journal of the Royal Statistical Society, series B* **39**, 1–38.

Diggle, P.J., Zeger, S.L. and Liang, K.-Y. (1994). *Analysis of Longitudinal Data*. Clarendon Press, Oxford.

Finney, D.J. (1971). *Probit Analysis*, 3rd ed. Cambridge University Press, Cambridge.

Guth, D.J., Jarabek, A.M., Wymer, L. and Hertzberg, R.C. (1991). Evaluation of risk assessment methods for short-term inhalation exposure. *Proceedings of the Air and Waste Management Association 84th Annal Meeting*, Paper No. 91-173.2. Vancouver, British Columbia.

Guth, D.J. and Raymond, T.S. (1996). A database designed to support dose-response analysis and risk assessment. *Toxicology*, **114**, 81–90.

Guth, D.J., Simpson, D.G., Carroll, R.J. and Zhou, H. (1997). Categorical regression analysis of acute inhalation exposure to tetrachloroethylene. *Risk Analysis* **17**, 321–332.

Hertzberg, R.C. (1989). Fitting a model to categorical response data with application to species extrapolation of toxicity. *Health Physics* **57**, 405–409.

Hertzberg, R.C. and Miller, M. (1985). A statistical model for species extrapolation using categorical response data. *Toxicology and Industrial Health* **1**, 43–63.

Jarabek, A.M. (1995). Consideration of temporal toxicity challenges current default assumptions. *Inhalation Toxicology* **7**, 927–946.

Lehmann, E.L. (1983). *Theory of Point Estimation*. Wiley, New York.

Li, B. (1996). A minimax approach to consistency and efficiency for estimating equations. *Annals of Statistics* **24**, 1283–1297.

McCullagh, P. (1980). Regression models for ordinal data. *Journal of the Royal Statistical Society, series B* **42**, 109–142.

McCullagh, P. and Nelder, J.A. (1989). *Generalized Linear Models*, 2nd ed. Chapman & Hall, London.

Morgan, B.J.T. (1992). *Analysis of Quantal Response Data*. Chapman Hall, London.

National Research Council (1993). *Guidelines for Developing Community Emergency Exposure Levels for Hazardous Substances*. National Academy Press, Washington, DC.

Ramsey, J.C. and Andersen, M.E. (1984). A physiologically based description of the inhalation pharmacokinetics of styrene in rats and humans. *Toxicology and Applied Pharmacology* **73**, 159–175.

Rusch, G.M. (1993). The history and development of emergency response planing guidelines. *Journal of Hazardous Materials* **33**, 193–202.

Simpson. D.G, Carroll, R.J., Schmedieche, H. and Xie, M. (1995). Documentation for categorical regression risk assessment (CatReg). Report to U.S. EPA, National Center for Environmental Assessment, Research Triangle Park, NC.

Simpson, D.G., Carroll, R.J., Xie, M. and Guth, D.L. (1996a). Weighted logistic regression and robust analysis of diverse toxicology data. *Communications in Statistics - Theory and Methods* **23**, 2615–2632.

Simpson, D.G., Carroll, R.J., Zhou, H. and Guth, D.L. (1996b). Interval censoring and marginal analysis in ordinal regression. *Journal of Agricultural, Biological and Environmental Statistics* **1**, 354–376.

ten Berger, W.F., Zwart, A. and Appelman, L. (1986). Concentration–time mortality response relationship of irritant and systemically acting vapors and gases. *Journal of Hazardous Materials* **13**, 301–309.

U.S. Congress. (1990). *Clean Air Act Amendments of 1990, PL 101-549*. November 15, 1990. U.S. Government Printing Office, Washington, DC.

U.S. Environmental Protection Agency (1991). Guidelines for developmental toxicity risk assessment; Notice. *Federal Register* **56**, 63798–63826.

U.S. Environmental Protection Agency (1995). The use of the benchmark dose approach in health risk assessment. Technical Report EPA/630/R-94/007. Environmental Protection Agency, Office of Research and Development, Washington, DC.

Xie, M. (1996). *Regression Modeling: Latent Structure, Theories and Algorithms*. Ph.D. dissertation, University of Illinois, Urbana-Champaign.

Xie, M. and Simpson, D.G. (1998). Regression modeling of ordinal data with non-zero baselines. *Biometrics*, **55** (to appear).

Xie, M., Simpson, D.G. and Carroll, R.G. (1997). Scaled link functions for heterogeneous ordinal response data. *Modelling Longitudinal and Spatial Correlated Data: Methods, Applications and Future Directions*, T.G. Gregoire, D.R. Brillinger, P.J. Diggle, E. Russek-Cohen, W.G. Warren and R.D. Wolfinger (eds.). Springer-Verlag, New York, pp. 23–36.

Zeger, S.L. and Liang, K.-Y. (1986). Longitudinal data analysis for discrete and continuous outcomes. *Biometrics* **42**, 121–130.

Workshop: Statistical Methods for Combining Environmental Information

Lawrence H. Cox
U.S. Environmental Protection Agency

1 The NISS-USEPA Workshop Series

Primary objectives of the NISS-USEPA cooperative research agreement were to identify important environmental problems to which statistical science could contribute, to perform interdisciplinary research on these problems and stimulate related research and problem identification within the broader statistical community, to assess important examples and areas of environmetric research, and to identify new research problems and directions. To provide a forum for identifying and examining new research and problem areas, a NISS-USEPA workshop series was established within the cooperative research program.

The workshop series comprised four independent annual workshops over the four-year cooperative research agreement. Workshop topics were drawn from broad statistical issues in environmental settings and motivated by current events and concerns in environmental science and management. The workshops were

- *Statistical Methods for Combining Environmental Information*
- *Spatial Sampling for the Environment*
- *Statistical Issues in Mechanistic Modelling for Risk Assessment*
- *Statistical Issues in Setting Air Quality Standards*

NISS Executive Director Jerome Sacks and U.S. EPA Senior Mathematical Statistician Lawrence Cox jointly selected workshop topics, planned the workshops, and selected the invited workshop presenters, discussants, and participants. Walter Piegorsch was instrumental in summarizing the first workshop, as were Dennis Cox and Kathy Ensor for the second workshop. G.P. Patil, Christopher Portier, and David Guinnup made invaluable contributions to the programs of the second, third and fourth workshops, respectively.

Workshop programs appear in Appendix C. NISS Technical Reports 12 and 38 summarize the 1993 and 1994 workshops, respectively, and NISS Technical Report 27, Cox and Piegorsch (1996), and Piegorsch and Cox (1996) resulted from the 1993 workshop, and Cox *et al.* (1997) resulted from the 1994 workshop. Sufficient time was not available under the cooperative agreement to summarize the 1995 and 1996 workshops. The 1996 workshop led to new research projects in the United States and Europe.

The remainder of this chapter is devoted to summarizing the first workshop on statistical methods for combining environmental information, giving some additional types of problems and case studies not addressed in earlier chapters of this book.

The reader is referred to the *web companion* for specific data sets and software that are related to the case studies in this chapter.

2 Combining Environmental Information

Environmental data collection is a difficult and expensive enterprise, necessitating reuse of available data, often for purposes unintended at the time the original data were collected. Data necessary to address even a single environmental issue often need to be compiled from several sources involving different variables, measurements, time and spatial scales, accuracy, precision, and completeness. This is a particularly difficult problem in environmental assessment where data and results from many diverse studies need to be evaluated and combined appropriately, e.g., from toxicological or epidemiological studies aimed at improving understanding of relationships between environmental conditions and health effects. Often, design criteria for data collection are not statistically based, so that important design and selection information is incomplete or unavailable. This may be true, e.g., for ecological studies that are not probability based. In addition, selection bias is often present but difficult to quantify. Data validation can be difficult and complex, involving piecemeal cross-comparisons among several data sources. Environmental data sets are often large and, consequently, difficult and expensive to manipulate and analyze. As demonstrated in this chapter, these problems pose important challenges to statistical science, and statistical methods are central to their solution.

The problem of combining environmental information arises in all aspects of environmental science. Data combination can be performed to improve accuracy or precision of estimates, to investigate relationships between data sets collected in different places, times, or ways, or to validate findings. The scope of environmental issues and the consequences of attendant remediation decisions involve the combination and cross-comparison of information from multiple sources and types. Motivated by these issues and immediate concerns for data combination arising from the U.S. EPA Environmental Monitoring and Assessment Program (EMAP), the passive smoking issue, and the need for reliable methods to combine the outcomes of toxicological studies for assessment purposes (discussed in Chapter 7), the first NISS-USEPA workshop was organized around the topic *combining environmental information*. Aspects of this topic were examined at the 1993 workshop using applications drawn from environmental epidemiology, assessment, and monitoring. Selected applications and findings of the workshop and subsequent cooperative research are presented below. Complete details of the workshop are reported in Cox and Piegorsch (1994). Cox and Piegorsch (1996) and Piegorsch and Cox (1996) offer more intensive analysis and additional examples.

The examples presented in the next section serve as case studies for combining environmental information.

3 Combining Environmental Epidemiology Information

Human or ecological health effects related to environmental conditions are often assessed by combining the results of designed experiments or epidemiological studies. Chapter 7 deals with combining designed toxicological experiments. The subject of combining epidemiological data and studies is examined in this chapter through two case studies. The first study was motivated by environmental tobacco smoke, a.k.a. passive smoking. This was among the leading environmental issues of the early 1990s. The central question was whether exposure to environmental tobacco smoke leads to negative health effects, e.g., for nonsmokers. A 1992 U.S. EPA study answered this question in the affirmative, leading to new, strict nonsmoking regulations in public buildings. Statistical evidence for this regulatory decision was discussed at the workshop. A second case study, on assessing respiratory effects on children of nitrogen dioxide exposure, was also discussed. The common statistical thread between these case studies is meta-analysis. The most simple form of meta-analysis is to combine P-values from several independent studies into a single P-value for the collection of studies.

The outcome of a quantitative study is often summarized by one or more statistics, e.g., scientific studies are frequently designed around statistical tests of hypothesis and summarized by corresponding P-values. Approaches that combine P-values, such as Fisher's method (viz., for c independent studies with corresponding P-values p_i, compare $\chi_F^2 = -2\sum_{i=1}^{c} \log(p_i)$ to a χ^2 reference distribution with 2c degrees of freedom, see Hedges and Olkin (1985)), synthesize disparate information into a single inference about the environmental phenomenon under study. Unless the data combination problem is fairly simple however, P-value combination may overlook relevant scientific differences among the information sources. An extension of P-value combination is effect size estimation, which combines a set of summary statistics to produce, e.g., a summary correlation coefficient or χ^2 statistic (Hedges and Olkin 1985, Ch. 5). The typical situation is to combine effect size estimates $(y_e - y_c)/\sigma_{e,c}$ across independent controlled experiments (y_e and y_c are estimates of the parameter of interest from experimental and control groups, and $\sigma_{e,c}$ is a pooled estimate of variance).

The first study combines simple P-value combination with statistical modeling. The second illustrates going beyond simple P-value combination through use of hierarchical regression.

3.1 Passive Smoking

For many years, there was concern that nonsmokers, as well as smokers, were at risk from exposure to "other people's" smoking. This became known as "passive smoking." A number of epidemiological studies were conducted in the United States and elsewhere to study passive smoking. Individually, these studies produced different estimates of risk and, in total, seemed to paint an ambiguous picture. To investigate this issue further, the U.S. EPA undertook to combine these studies using statistical (meta-analytical) techniques. Using these methods, the results were not ambiguous and indicated that passive smoking does lead to a quantifiable increase risk of cancer death. These findings led to a series of now familiar federal and state laws and regulations restricting smoking in public buildings, restaurants, etc. In addition, emerging studies appear to relate passive smoking to heart disease.

An important example of characterizing risk from multiple data sources is illustrated by a U.S. EPA study (1992). The study was aimed at combination of epidemiologic data on the health effects of environmental tobacco (or "passive") smoke. Thirty epidemiologic studies were considered as part of the EPA analysis. In all cases, the measured effect was the relative increase in risk for lung cancer mortality over that risk for non-exposed controls, i.e., the relative risk of exposure death to unexposed death: $RR = \Pr\{D|E_+\}/\Pr\{D|E_-\}$. When the disease prevalence in the population is small, relative risk may be estimated via the corresponding ratio of odds of exposure for cancer deaths ("cases") to odds of exposure for controls (Breslow and Day 1980, Sec. 2.8). Statistical methods are then employed to test whether the odds ratio equals 1. (Substitution of odds ratios is a standard technique in epidemiology that is also used in the second case study.)

For the 30 environmental tobacco smoke (ETS) studies, the "exposed" groups were female nonsmokers whose spouses smoked. The estimated relative risks for death due to lung cancer ranged from 0.68 to 2.55 (see Table 1). When analyzed separately, only one of the eleven U.S.-based studies showed a significant increase ($P=0.03$) in the odds of lung cancer mortality after ETS exposure. At issue was whether proper combination of the individual odds ratios would identify an overall increased risk of lung cancer death due to ETS.

The U.S. EPA investigators went beyond simple P-value combination. As part of the data combination, relative risk models were adjusted for background exposures believed to decrease the observed (unadjusted) risk. For example, women without spouses who smoke are still exposed to background ETS, via workplace or public/outside-the-home exposures. Thus, the unexposed group may still exhibit some background lung cancer risk due to ETS, above and

Region	Study	Estimated RR_i	90% Confidence Limits	Weight w_i
G	kala	1.92	(1.13, 3.23)	1.98
G	tric	2.08	(1.31, 3.29)	12.76
HK	chan	0.74	(0.47, 1.17)	13.01
HK	koo	1.54	(0.98, 2.43)	13.12
HK	kamt	1.64	(1.21, 2.21)	29.83
HK	lamw	2.51	(1.49, 4.23)	9.94
J	akib	1.50	(1.00, 2.50)	12.89
J	hiraCoh	1.37	(1.02, 1.86)	29.98
J	inou	2.55	(0.90, 7.20)	2.50
J	shim	1.07	(0.70, 1.67)	14.32
J	sobu	1.57	(1.13, 2.15)	26.16
USA	brow	1.50	(0.48, 4.72)	2.07
USA	buff	0.68	(0.32, 1.41)	4.92
USA	butlCoh	2.01	(0.61, 6.73)	1.88
USA	corr	1.89	(0.85, 4.14)	4.32
USA	font	1.28	(1.03, 1.60)	55.79
USA	garf	1.27	(0.91, 1.79)	23.65
USA	garfCoh	1.16	(0.89, 1.52)	37.78
USA	humb	2.00	(0.83, 4.97)	3.38
USA	jane	0.79	(0.52, 1.17)	16.46
USA	kaba	0.73	(0.27, 1.89)	2.86
USA	wu	1.32	(0.59, 2.93)	4.21
EU	holeCoh	1.97	(0.34, 11.67)	0.87
EU	lee	1.01	(0.47, 2.15)	4.68
EU	pers	1.17	(0.75, 1.87)	12.97
EU	sven	1.20	(0.63, 2.36)	6.21
C	gao	1.19	(0.87, 1.62)	28.00
C	geng	2.16	(1.21, 3.84)	8.12
C	liu	0.77	(0.35, 1.68)	4.40
C	wuwi	0.78	(0.63, 0.96)	60.99

TABLE 1. Summary relative risk information for 30 individual studies of lung cancer risk after ETS exposure, grouped by geographic region. Region codes: G=Greece, HK=Hong Kong, J=Japan, USA, EU=Western Europe, C=China. Study abbreviation codes are from EPA (1992, Table 5.9) where Coh=cohort study. Estimated relative risk RR_i is adjusted for smoker misclassification. Weight $w_i = 1/\text{Var}[\log(\widehat{RR}_i)]$.

beyond an idealized "baseline" group that received no ETS exposure whatsoever.

The following model was developed to perform the adjustment. The baseline risk to an idealized group with no ETS exposure was taken as 1. Next, risk to the unexposed group was modeled via $1 + \beta_i d_i$, where β is the increased risk per unit dose and d is the mean dose level in the unexposed group. Risk to the "exposed" group was modeled as $1 + z_i\beta_i d_i$, where z is the ratio between the mean dose level in the exposed group and the mean dose level in the unexposed group. This gives $z_i > RR_i > 1$, where $RR_i = (1 + z_i\beta_i d_i)/(1 + \beta_i d_i)$ is the "observed" relative risk. Under this model, the adjusted risk for the exposed group relative to the baseline group is the quantity of interest. It is calculated as the ratio of the "exposed" group risk to the baseline risk. As the baseline risk equals 1 under this model, then $RR_i^* = 1 + z_i\beta_i d_i$. Over the 30 studies, these adjusted estimates are pooled, so that, on a logarithmic scale,

$$\log(\widehat{RR}^*_{Pooled}) = \frac{\sum w_i \log(\widehat{RR}^*_i)}{\sum w_i},$$

where the per-study weights are $w_i = 1/\text{Var}[\log(\widehat{RR}^*_i)]$. Using only the observed relative risks, combination of all the U.S. studies in the EPA analysis produced a statistically significant ($P=0.02$) estimate for increased lung cancer risk of 1.19. Adjustment for background exposures, however, gives stronger evidence: The relative risk estimate increases to a value of $\widehat{RR}^*_i = 1.59$, i.e., an estimated 59% increase in lung cancer mortality in U.S. non smokers when exposed to environmental tobacco smoke.

In this example, relatively simple statistical methods were sufficient for the data combination. This is often not the case. These results are encouraging and illustrate the need for ongoing research into trenchant meta-analysis methods in environmental science and increased use of such methods. Environmental and public health policy decisions involve serious economic and social outcomes. These decisions should be based on the most reliable and complete combination of available information.

3.2 Nitrogen Dioxide Exposure

Meta-Analysis of Nitrogen Dioxide Data

Quantifying risk of respiratory damage after exposure to airborne toxins is an ongoing concern in modern environmental epidemiology. The next case study involves an EPA meta-analysis of respiratory damage after indoor exposure to nitrogen dioxide, NO_2. Previous studies had produced mixed results on the risk of NO_2 exposure. Of particular concern is the risk for children, who respirate 50% more air by body weight than adults. The EPA study combined information on the relationship of NO_2 exposure to respiratory illness from these separate studies.

Using as an outcome variable the presence of adverse lower respiratory symptoms in children aged 5 to 12 years, odds ratios were employed to estimate the relative risk (RR) for increased lower respiratory distress in exposed population(s). A set of nine North American and western European studies reported odds ratios ranging from 0.75 to 1.49. Separately, only four of the nine odds

Study Code	Estimated RR	90% Confidence Limits	Weight w_i
M77	1.31	(1.18, 1.45)	258.90
M79	1.24	(1.11, 1.39)	219.67
M80	1.53	(1.11, 2.11)	26.10
M82	1.11	(0.87, 1.42)	44.88
W84	1.08	(0.99, 1.15)	489.33
N91	1.47	(1.21, 1.79)	71.50
E83	1.10	(0.83, 1.45)	35.17
D90	0.94	(0.70, 1.26)	31.30
K79	1.10	(0.78, 1.52)	25.03

TABLE 2. Summary relative risk information for nine individual studies of childhood respiratory disease risk after NO_2 exposure. Study abbreviation codes adapted from DuMouchel (1994). Weight $w_i = 1/\text{Var}[\log(\widehat{RR})]$.

ratios suggested a significant increase in respiratory distress due to NO_2 exposure (see Table 2). To combine the observed odds ratios, inverse variance weighting was applied: Separate estimates of the parameter of interest and its variance were computed, each estimate was weighed inversely to its estimated variance, and the weighted estimates were added together for a final estimate. For these data, this led to a combined RR estimate of 1.18, with 95% confidence limits from 1.11 to 1.25; that is, the meta-analysis suggested that increased NO_2 exposure can lead to an increased risk of respiratory illness of about 11–25% over unexposed controls (Hasselblad et al. 1992).

Hierarchical Bayes Meta-analysis of Nitrogen Dioxide Data

The work of Hasselblad et al. (1992) was presented in one of the workshop sessions. The discussant for the session was William DuMouchel. DuMouchel took the nine studies in the EPA NO_2 analysis and applied a hierarchical model for the separate log odds ratios. Prior normal distributions were placed on the underlying mean log odds ratios, and it was assumed that these mean log odds ratios were themselves normally distributed. The model extended the hierarchy by placing additional hierarchical distributions on the hyperparameters of the normal prior distributions. A further description of the model and the hierarchical approach is available in DuMouchel (1994).

Although complex, this hierarchical model was easily manipulated to provide posterior point estimates and posterior standard errors for the log odds ratios. These estimates were then employed to find an overall, combined estimate of relative risk via inverse variance weighting (as in Section 3). Applied to the nine NO_2 studies, the resulting combined (posterior) log(RR) is 0.1614 (odds ratio $e^{0.1014} = 1.175$), buttressing the EPA estimate.

DuMouchel (1994) also described a hierarchical regression model that adjusted the log(RR) estimate for important covariates such as household smoking and gender. Applied to the nine NO_2 studies, five of the nine studies exhibited significant ($P < 0.05$) posterior increases in relative risk, yielding a covariate-adjusted combined log(RR) of 0.1567 (odds ratio $e^{0.1567} = 1.170$). Data combination that incorporates important covariates in hierarchical Bayes analyses is a straightforward extension of the simple hierarchical model. Combining the similar information sources with prior distribution(s) via a regression relationship can, in effect, smooth out instabilities or other outlying features in the data. This leads to posterior estimates that can outperform those from other, nonhierarchical approaches (Greenland 1994). The techniques illustrated in the DuMouchel analysis have become more widely used since 1993.

A useful by-product of the hierarchical Bayes approach is that it provides an updated estimate of the between-study mean and of the distribution of the individual study outcomes (in this case, estimates of log(RR)) through "borrowing strength" and "shrinkage." This information can be used to assess the studies individually and as a group. For example, are there subgroups or strata of studies? On what features of the study might the stratification be based? Do some studies appear anomalous in the context of the complete set of studies? Are there reasons to disregard these studies? Or, is there a basis to differentially weight the studies for combination? The statistical considerations have the potential to increase understanding of the underlying scientific problems and assessment of the individual study outcomes.

4 Combining Environmental Assessment Information

The examples presented in this section illustrate statistical tools available for combining environmental information.

4.1 A Benthic Index for the Chesapeake Bay

Environmental information is complex and multidimensional. Tools for environmental assessment need to provide an accurate portrayal of status and trends in environmental condition, and at the same time be simple enough to be useful in policy planning and as public information. What is sought are environmental indexes. Similar to economic indexes such as the Consumer Price Index, indexes measuring unemployment, cost of new construction, etc., environmental indexes would combine quantitative information (typically, individual environmental indicators such as relating to the health or abundance of a particular species, or the concentration of a particular compound such as phosphorus or dissolved oxygen) into a single value and represent that value on a simple, interpretable scale. In this subsection, we present a case study of the development of an environmental index measuring estuarine condition.

In the early 1990s, the U.S. EPA and the State of Maryland conducted an assessment of the effectiveness of environmental restoration in the Chesapeake Bay. A variety of ecological approaches, in the form of ecological indexes, are available to measure estuarine health, based typically on what is in the water (algae, chemicals, etc.), what lives in the water (fish and marine animals), or what habitats the estuarine sediment. Of concern in this study was how nutrient abundance affects the plant and animal communities (benthos) in bay sediment, and therefore an ecological nutrient index (benthic index) for the assessment was developed by estuarine ecologists and biologists. Based on their expert knowledge, an index represented as a weighted sum of quantifiers related to individual species was selected:

$$z = w_1 x_1 + w_2 x_2 + w_3 x_3 + w_4 x_4 + w_5 x_5.$$

The quantifiers are as follows: x_1, the salinity-adjusted number of species; x_2, the percent of total benthic bivalve abundance; x_3, the number of amphipods (crustaceans with multipurpose feet); x_4, the average weight per polychaete (a type of segmented marine worm); and x_5, the number of capitellids (a special form of polychaete) observed at each location. The w_j are the weights, which must be determined.

Numerous samples were collected over four years from 31 sample sites in Chesapeake Bay. The study was not based on a probability sampling design. Rather, site selection was based on ecological criteria. To estimate the weights, experts assigned index values to each site at multiple time points. These values, together with site-based measurements of the five index parameters taken at corresponding times, were used to estimate the index weights w_i via regression. In effect, regression was used to combine the expert judgment in the form of a weighted linear combination of the species-specific indicators. This resulted in a formula for the benthic index:

$$z = .011 x_1 + .671 x_2 + .817 x_3 + .577 x_4 + .465 x_5.$$

A goal was to represent the benthic condition of the Bay as a map of the z index. From this, other assessments could be made, e.g., estimating the percentage of Bay acreage exhibiting degraded biotic conditions or identifying acreage in need of restoration. For this, the index must be predicted at unobserved locations. This was accomplished by modeling the index in two parts: using regression based on location and depth to account for large-scale variation in the z index; and then Kriging the regression residuals to account for small-scale variation. Alternatively, the x values could have been Kriged and corresponding predicted z values obtained from the index formula, but presumably the index is less variable spatially than its individual component parts. Another alternative would have been to include the location and depth information in the Kriging and base prediction solely on spatial interpolation via Kriging.

4.2 Hazardous Waste Site Characterization

Soil or water samples are collected at multiple locations called sampling sites within a hazardous waste site to determine whether environmental cleanup is necessary has been successful, or needs to be continued. This determination is made based on estimates of remaining contamination obtained by combining site measurements. Fisher's method for P-values combining from c independent studies can be used to combine environmental data for site characterization, as follows. Assume that soil samples containing concentrations of c toxic chemicals are taken at each of k locations within a hazardous waste site, that the k location-specific sampling distributions are i.i.d. with known mean and variance vectors, and that the c chemical concentrations, C_{ij}, at each location, j, are independent. Then, independent F-statistics, F_i, with corresponding P-values, P_i, $i = 1, ..., c$, can be used to test hypotheses of combined effective cleanup across the sites for each chemical. Fisher's method is then used as an omnibus test of combined effective cleanup across all chemicals and sites by combining the P_i into χ_F^2.

However, in realistic situations samples are not i.i.d. or have unknown variances, and Fisher's method may be inapplicable or suboptimal. An extended method due to Mathew et al. (1993) and presented at the workshop will be illustrated for the case $k = 2$. Assume that the site-specific sampling distributions are not i.i.d. with known mean and variance. Compute the two-sample between-chemical correlation coefficient R and its P-value P_R relative to an omnibus null hypothesis, $H_0 : C \leq C_0$, of effective cleanup across both sites (viz., the actual concentration of each chemical at each site is below the contamination threshold, C_{0j}, for that chemical). Under H_0, F_1, F_2, and R are independent, and therefore the test statistic $\chi_{12}^2 = -2(\log(P_1) + \log(P_2) + \log(P_R))$ can be compared to a reference χ^2 distribution with 6 df.

Often, there is correlation among chemical concentrations at a sampling location, and it is inappropriate to combine P-values directly. Solutions are needed for this problem.

4.3 Estimating Snow Water Equivalent

A problem of importance to the National Weather Service is to estimate the amount of water to be released into streams and lakes during the spring thaw from accumulated mountain snow. This is referred to as the snow water equivalent (SWE) problem. investigated by Carroll et al. (1995). Data were available from two sources: data collected from terrestrial sites and data from airborne monitoring. In the terminology of spatial statistics, these two data sets exhibit different spatial support, viz., point support and areal support, respectively. The class of such "change of support" problems presents challenging problems in spatial statistics.

One approach to the data combination would be to compute separate kriging estimates from the terrestrial and airborne data sets and combine these estimates, e.g., using inverse variance weighting. However, direct data combination, where possible, is capable of producing superior results. Carroll *et al.* (1995) did precisely that. They solve the SWE problem by direct computation of the combined terrestrial–airborne data covariance matrix, and produce a combined kriging estimate. This approach can be generalized to other situations, viz., if the two types of support are nonoverlapping. The problem of combining large- and small-scale spatial data is one of many outstanding problems in environmental data combination, and arises frequently. Change of support is a useful approach to some of these problems, but can be challenging technically (see Myers (1993) on change of support).

5 Combining Environmental Monitoring Data

Environmental monitoring data are available from many sources, ranging in scope from single ponds or hazardous waste sites to national ambient air monitoring networks, over different environmental media (air, water, soil), and for different, often overlapping, time and spatial scales. Some environmental monitoring programs, such as EMAP and the U.S. National Resources Inventory (NRI), are based on probability sample (P-sample) designs.

For purposes here, a P-sample design is one for which each potential sampling location is assigned a non-zero probability of being selected for data collection; a sample of locations (typically of predetermined size) is selected according to this probability structure; and data (environmental samples) are collected at these locations according to established data collection protocols. Some environmental surveys (e.g., EMAP and NRI) and studies involve P-samples. However, the vast majority do not. We refer to data not collected according to a P-sampling design as NP-samples. Thompson (1992) covers selected methods in modern probability sampling.

Environmental monitoring data are difficult and expensive to collect, and statistical methods are needed to combine monitoring data to maximize environmental understanding and effective environmental management. The familiar paradigms for data combination in ecological assessment are: combining two P-samples; combining a P-sample with an NP-sample; and combining two NP-samples. Emerging consideration is also given to additional, e.g., spatial, structure inherent in the data.

The examples presented in this section illustrate statistical methods for combining environmental information.

5.1 Combining P-Samples

In 1993, EMAP was emerging as the nation's largest environmental assessment program and was to be based on a probability-sample design. The issue of combining P-samples was on the forefront of ecological assessment.

Three standard approaches are available for combining data or estimates from two P-samples A and B. The first combines weighted estimates from the separate P-samples. The most general method is based on inverse variance weighting (discussed in Section 3). This produces an unbiased minimum variance combined estimate

$$\hat{Y} = \frac{\frac{\hat{Y}_A}{\sigma_A^2} + \frac{\hat{Y}_B}{\sigma_B^2}}{\frac{1}{\sigma_A^2} + \frac{1}{\sigma_B^2}}$$

(Hartley 1974). This method is also applicable to NP-samples.

A more trenchant approach for P-samples is dual frame estimation which produces an unbiased minimum variance combined estimate of a population total Y. Let A^* and B^* denote the population frames corresponding to the samples A and B, and consider the three nonoverlapping subsamples of $A \cup B$, $A \cap B$, $A \setminus B$ (viz., units in A not contained in B), and $B \setminus A$. A family of unbiased combined estimates of Y is obtained from

$$\hat{Y}_q = \sum_{i \in A \setminus B} (1/p_i^{A \setminus B}) y_i^{A \setminus B} + \sum_{i \in B \setminus A} (1/p_i^{B \setminus A}) y_i^{B \setminus A}$$
$$+ q \sum_{i \in A \cap A^* \cap B^*} (1/p_i^{A \cap B}) y_i^{A \cap B} + (1-q) \sum_{j \in B \cap A^* \cap B^*} (1/p_j^{A \cap B}) y_j^{A \cap B}$$

for sample observations and probabilities of selection y and p and for $0 \leq q \leq 1$. Optimizing over q yields a minimum variance unbiased combined estimate (Hartley 1974).

The second method for combining two P-samples is based on Post-stratification. Strata are defined using shared frame attributes or subsamples that partition the two samples. Both samples are poststratified, namely, sample unit weights are revised proportional to stratum size (as above) and used to compute revised estimates for the parameter of interest.

Better results are expected if samples, rather than estimates, are combined. The third method combines two P-samples into one P-sample, based on combined-sample inclusion probabilities for all units, usually computed from their first- and second-order inclusion probabilities in the original samples. Unfortunately, such information is not always available. Also, because this method depends on the degree to which the population frames coincide, analysis can be restricted to the intersection of the frames, or each sample might be augmented with only some but not all units from the other sample.

Combining P-samples is the best developed area of statistical data combination methodology, but due to incomplete information, it can be subject to limitations as described above. Emerging methods discussed at the workshop involve adaptive sampling and ranked set sampling. Adaptive sampling uses information collected on the first k sample units to decide where to sample the $(k+1)$st unit (viz., when using random grid sampling to estimate population size for a plant or animal species, whenever the species is detected in one sample grid, add the surrounding grids to the sample). Ranked set sampling employs (subjective) assessments based on a covariate to prestratify units (e.g., if the variable of interest, say vegetation, is believed to be related to forest density, stratify potential sample sites into high, medium, and low categories based on a visual assessment of density, then draw the sample based on this stratification). Both methods combine auxiliary data with a default sampling method (such as simple random sampling) with the aim of improving sample efficiency (Thompson 1992, Part VI). A related technique uses covariate adjustment based on an existing P-sample at the time of selecting a second P-sample to improve the efficiency of the second sample (Patil 1996).

5.2   Combining P- and NP-Samples

Overton et al. (1993) address the problem of combining a P-sample with an NP-sample. The NP-sample is identified with a subset of the population represented by the P-sample, such as by clustering based on NP-sample attributes. The population is partitioned so that the NP-sample is a representative sample of one partition, reducing the problem to that of combining two P-samples. This method is imperfect, however: Representativeness is liable to be difficult to verify, and, if false, is liable to introduce systematic bias into combined sample estimates. An alternative is to include the NP-sample in the combined sample as a separate stratum of self-representing units; i.e., assign probability of selection one to each NP-unit. Unfortunately, this will not improve precision unless the original P-sample is small. Needed are further methods that use structure and data from a P-sample to enhance representativeness of models based on NP-samples.

5.3   Combining NP-Samples

A regression method for augmenting P-samples (Overton et al. 1993) can be used to augment NP-samples. Let y denote a variable on the first sample but not the second sample; x denotes variables common to both samples. Regress y on x in the first sample, $y = x\beta + \epsilon$, and apply this regression equation to x variables on the second sample to predict y for each second sample unit, $\hat{y}_0 = x_0\hat{\beta}$. Problems with this method include failure of the regression to account for true variation in y and potentially undetected bias in the regression due to first-sample selection bias.

5.4   Combining NP-Samples Exhibiting More Than Purposive Structure

Plant, wildlife, and other ecological sampling is often based on a period of observation at preselected sampling locations. The resulting data are likely to be nonrandom, due to censoring mechanisms such as bias toward observing larger over smaller specimens or biases based on direction, terrain, etc. Observed distributions need to be weighted to account for observer-observed bias. For example, if each observation v has probability $1 - w(v)$ of not being observed, then the observed probability density function (pdf) is the true pdf weighted by $w(v)$, and observed data are weighted by $1/w$ prior to estimation. Presence of such bias is often signaled by overdispersion, viz., excessive variance due, e.g., to undetected bias or nonconstant mean.

Patil (1991) models overdispersion using double exponential distributions

$$f_{\mu,\theta}(y) = c(\mu, \theta)\theta^{1/2}(f_\mu(y))^\theta (f_y(y))^{1-\theta},$$

where $f_\mu(y) = \exp\{yA(\mu) - B(\mu) + D(y)\}$ is the linear exponential family with mean μ and variance $1/A'(\mu)$ with $\mu A'(\mu) - B'(\mu) = 0$. The double exponential family enjoys exponential family properties for both mean and dispersion parameters, enabling application of common regression methods. Overdispersion is common in environmental data and may be due, for example, to clumped sampling, heterogeneity, or selection bias. Patil connects double exponential families and weighted distribution functions via

$$f_{\mu,\theta}(y) = w_{\mu,\theta}(y) f_\mu(y) / E[w_{\mu,\theta}(Y)]$$

with

$$w_{\mu,\theta}(y) = (f_y(y)/f_\mu(y))^{1-\theta},$$

enabling bias reduction through modeling of overdispersion. If the observed data are P-sample data, then they can be treated as a (weighted) two-stage P-sample and standard P-sample methods can be applied. If not, by accounting for some bias, weighting has improved the representativeness of observed data, and standard methods can be applied to the weighted data. Weighted distribution functions add the ability to combine empirical pdf's with probability distributions of observer bias prior to analysis and to eliminate or reduce bias in overdispersed data. Extensions of this approach to multivariate environmental data are needed.

6 Future Directions

Several open research problems and directions for research were described in the preceding examples. Problems which promise to contribute significantly both to environmental understanding and to statistical methodology include the development of methods for combining NP-sample data, development of a theoretical framework for integrating spatial, and P-sample methods for environmental assessment, new methods and extensions of existing methods for combining spatial data collected at different aggregation scales, modeling approaches that eliminate or reduce bias in environmental data, and extensions of meta-analytic methods to the environmental arena, including methods for combining correlated studies such as involving different contaminants at the same sites, or multiple studies on the same data set, and hierarchical methods that enable combination and intercomparison of different environmental studies.

References

Breslow, N. and Day, N. (1980). *Statistical Methods in Cancer Research. I. The Analysis of Case-Control Studies*, Vol. 32. IARC Scientific Publications, Lyon, France.

Carroll, S., Day, G., Cressie, N. and Carroll, T. (1995). Spatial modeling of snow water equivalent using airborne and ground-based snow data. *Environmetrics* **6**, 127–139.

Cox, D., Cox, L. and Ensor, K. (1997). Spatial sampling and the environment: Some issues and directions. *Environmental and Ecological Statistics* **4**, 219–233.

Cox, L.H. and Piegorsch, W.W. (1994). Combining environmental information: Environmetric research in ecological monitoring, epidemiology, toxicology, and environmental data reporting. Technical Report 12. National Institute of Statistical Sciences, Research Triangle Park, NC.

Cox, L.H. and Piegorsch, W.W. (1996). Combining environmental information I: Environmental monitoring, measurement and assessment. *Environmetrics* **7**, 299–308.

DuMouchel, W. (1994). Hierarchical Bayes linear models for meta-analysis. Technical Report 27. National Institute of Statistical Sciences, Research Triangle Park, NC.

Greenland, S. (1994). Hierarchical regression for epidemiologic analysis of multiple exposures. *Environmental Health Perspectives* **102**, Suppl. 8, 33–39.

Hartley, H. (1974). Multiple frame methodologies and selected applications. *Sankhya, series B* **36**, 99–118.

Hasselblad, V., Eddy, D.M. and Kotchmar, D.J. (1992). Synthesis of environmental evidence: Nitrogen dioxide epidemiology studies. *Journal of the Air & Waste Management Association* **42**, 662–671.

Hedges, L. and Olkin. I. (1985). *Statistical Methods for Meta-Analysis*. Academic Press, Orlando, FL.

Mathew, T., Sinha, B. and Zhou, L. (1993). Some statistical procedures for combining independent tests. *Journal of the American Statistical Association* **88**, 912–919.

Myers, D. (1993). Change of support and transformations. *Geostatistics for the Next Century*, R. Dimitrakopoulous (ed.). Kluwer Academic Publishers, Dordrecht, pp. 253–258.

Overton, J., Young, T. and Overton, W.S. (1993). Using found data to augment a probability sample: Procedure and a case study. *Environmental Monitoring and Assessment* **26**, 65–83.

Patil, G.P. (1991). Encountered data, statistical ecology, environmental statistics, and weighted distribution methods. *Environmetrics* **2**, 377–423.

Patil, G.P. (1996). Using covariate-directed sampling of EMAP hexagons to assess the statewide species richness of breeding birds in Pennsylvania. Technical Report 95-1102. Center for Statistical Ecology and Environmental Statistics, Pennsylvania State University, University Park, PA.

Patil, G.P., Sinha, A.K. and Taillie, C. (1994). Ranked set sampling. *Handbook of Statistics Volume 12: Environmental Statistics*, G.P. Patil and C.R. Rao (eds.). North-Holland, Amsterdam, pp. 103–166.

Piegorsch, W.W. and Cox, L.H. (1996). Combining environmental information II: Environmental epidemiology and toxicology. *Environmetrics* **7**, 309–324.

Thompson, S.K. (1992). *Sampling*. Wiley, New York.

U.S. Environmental Protection Agency (1992). Respiratory health effects of passive smoking: Lung cancer and other disorders. Technical Report 600/6-90/006F. Environmental Protection Agency, Washington, DC.

Appendix A: FUNFITS, Data Analysis and Statistical Tools for Estimating Functions

Douglas Nychka
North Carolina State University

Perry D. Haaland and Michael A. O'Connell
Becton Dickinson Research Center

Stephen Ellner
North Carolina State University

1 Introduction

FUNFITS is a suite of functions that enhance the S-PLUS^{TM} statistical package and facilitate curve and surface fitting. This software grew up as a project to augment S-PLUS with additional nonparametric fitting strategies. Part of this effort was to focus on three areas rich in applications of function estimation and inference: spatial statistics, response surface methodology, and nonlinear time series/dynamical systems. This activity has also led to methods for spatial sampling and design of experiments when the response is expected to be a complex surface. Implementation of these methods with an object-oriented approach gives a high level of integration within the FUNFITS package. This allows the models to be easily fit, summarized, and checked. The main methodological contributions of FUNFITS are

- Thin-plate splines
- Spatial process models
- Space-filling designs
- Neural network regression
- Response surface methodology using nonparametric regression
- Nonlinear autoregressive process models and the analysis of nonlinear dynamic systems

Where possible the statistical methods are supported by generic functions that provide predicted values and derivatives, diagnostics plots, and simple ways to visualize the functions using contour and surface plots. In addition, there are generic functions to optimize the estimated function and to find a path of steepest ascent. Some discipline has been kept to implement the methods in the S language; however, a few of the numerically intensive steps are completed using FORTRAN subroutines. All the source code is included in the distribution, so it is possible to examine the algorithms when needed. The emphasis on S code and the organization of the FORTRAN objects helps to make parts of the package easy to modify.

This appendix gives an overview of the FUNFITS package with some examples for methods particularly useful for spatial data. The reader is referred to the FUNFITS manual (Nychka et al. 1996) for a complete description of this package. Both the manual and the software can be accessed through the World Wide Web from the FUNFITS homepage (http://www.stat.ncsu.edu/~nychka/funfits) or through the NISS homepage (http://www.niss.org/).

2 What's So Special About FUNFITS?

This project was started with an interest in implementing surface- and curve-fitting methods that provide a global representation for the estimated functions. This form is complementary to local methods such as loess already supported by S-PLUS. By a global method we mean that the function has a specific (closed form) formula depending on a set of parameters, e.g.,

$$\hat{f}(x) = \sum_{k=1}^{M} c_k \psi_k(x),$$

where the $\{\psi_k\}$ are a set of basis functions and the $\{c_k\}$ are estimated parameters. Of course, the nonparametric flavor of this estimate means that M is similar in size to the number of data points and the estimates of the parameters are constrained in such a way that the resulting estimate is smooth. Global methods are especially useful for smaller sample sizes, low noise situations, or applications where some inferences are required.

Although the use of this package requires some knowledge of S-PLUS, many of the FUNFITS functions are easy to use and require a minimum of S-PLUS commands. Also, there are additional functions in FUNFITS that help to round the corners of what we perceive to be rough edges in S-PLUS. Faced with the unpleasant task of learning a new programming/data analysis language, the reader should keep in mind that no system will make function and curve fitting a trivial exercise. The data structures can be fairly complicated and some scrutiny of and interaction with the data are always required to prevent nonsensical results. Because S-PLUS already supports many statistical functions and has very flexible graphics, it is a natural platform for FUNFITS. For a gradual introduction to S-PLUS, the reader might consider S-Lab (Nychka and Boos 1997), a series of labs teaching data analysis and some of the some of basic functions of S-PLUS incrementally. The overview by Venables and Ripley (1994) is also helpful.

2.1 An Example

Another project goal is to use the object-oriented features of the S language to make the functions easier to use. To illustrate these ideas, here is an example of fitting a thin-plate spline surface to the summer 1987 ozone averages for 20 stations in the Chicago urban area. The data are in the S data set, ozone with the components: x a 20 × 2 matrix with approximate Cartesian coordinates[1] for the station locations, a vector y with the average ozone values for each station, and lon.lat, a matrix with the geographic locations of the stations. The text following the S-PLUS prompt: > is what has been typed in (comments follow the pound sign, #).

```
> ozone     # print the ozone data to the screen
> plot(ozone$lon.lat) #plot locations of stations
> US(add=T)       # with medium resolution US map overlaid
> ozone.tps<- tps(ozone$x, ozone$y)    # thin plate spline
                                       # fit to data
```

The results from fitting the thin-plate surface to the 20 mean ozone values are now stored in the output data set ozone.tps. Since a smoothing parameter has not been specified, the default is to estimate the amount of smoothness in the surface by cross-validation.

The next step is to look at a summary of the fit:

```
> summary(ozone.tps)

Call:
tps(x = ozone$x, y = ozone$y)

 Number of Observations:                          20
 Degree of polynomial null space ( base model):   1
 Number of parameters in the null space           3
 Effective degrees of freedom:                    4.5
 Residual degrees of freedom:                     15.5
 Root Mean Square Error:                          4.047
 Pure Error:                                      NA
 RMSE (PRESS):                                    4.4594
 GCV (Prediction error)                           21.17
 Mutiple R-squared:                               0.3521
 Adjusted R-squared:                              0.07126
 Q-squared:                                       -0.1868
 PRESS:                                           397.7244
```

---

[1] To make this example simpler and the smoothing comparable in the north-south and east-west directions, the geographic coordinates for each station have been converted to (approximate) X-Y rectangular locations in miles. See the section on spatial process estimates for an analysis of these data using lon/lat coordinates.

```
Log10(lambda):                                         -0.034
m, power, d, cost:                                     2, 2, 4, 1
Residuals:
    min   1st Q  median  3rd Q   max
  -6.712 -1.426 -0.4725  1.326  7.912
Method for specifying the smoothing parameter is  GCV
```

This summary indicates some basic features of the regression. With approximately 4.5 degrees of freedom estimated for the surface, one can expect that the fit is similar to a quadratic fit (five parameters) but must have more structure than just fitting a linear function.

To assess the quality of the fit one can view residual plots and the GCV function:

```
> plot(ozone.tps)
```

This graphical output is given in Figure 1. Based on the two plots of fitted values and residuals, we see that the estimate has some problems with overfitting ozone at the lower values. Good models tend to have small values of the cross-validation function, and we see on the lower plots that models with higher effective numbers of parameters (or equivalently smaller values of the smoothing parameter λ) also do not predict as well. One alternative in this case is to use the variances based on the daily ozone values as a way to determine the appropriate amount of smoothness.

One can also examine the fitted surface (see Figure 2):

```
> surface(ozone.tps,'Fitted ozone surface')
```

As expected from the summary, we see that the estimated surface is similar to a quadratic function in shape. The surface has been restricted to the convex hull of the stations and so, for example, it is not extrapolated over Lake Michigan (the upper right corner of the region).

Besides a plot of the fitted surface, it is also easy to determine a rough measure of uncertainty in the estimate. This involves two steps, first evaluating the standard errors at a grid of locations and then passing these evaluations to one of the surface plotting programs in S. Here is an example that gives the image plot in Figure 3.

```
> ozone.se <- predict.surface.se(ozone.tps)
> image(ozone.se, 'SE of fit')   # or use contour  or    persp
> points( ozone$x) # add in the station locations
 # Now add contours of the predicted spline fit
> contour( predict.surface( ozone.tps), add=T)
```

Thus, with a minimal amount of work, it is possible to fit, check, and interpret a fairly complicated nonparametric regression method. The main intent of FUNFITS is that functions such as surface or predict should work with a variety of methods. The burden is on the software to sort out salient differences

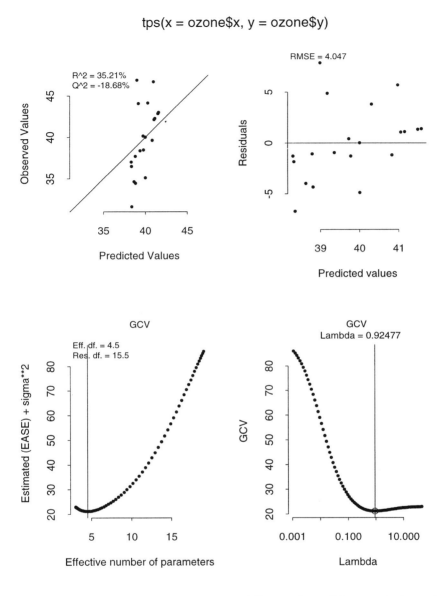

FIGURE 1. Summary plot of tps fit to the average daily ozone for 20 Chicago monitoring stations.

between the objects that are passed to the function and still produce sensible results. In this example, the ozone.tps object contains the description of the estimated spline function but the ozone.se object consists of standard errors evaluated on a (default) grid of points. In either case, a perspective or contour plot of the function makes sense and, thus, the surface function is designed to

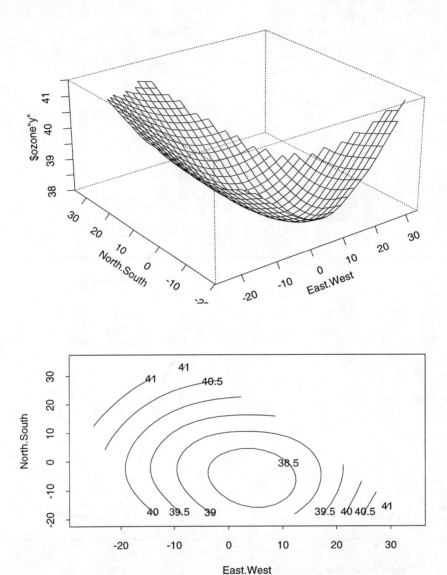

FIGURE 2. Average daily ozone surface for 20 Chicago monitoring stations.

handle both types of objects. Another important principle in this example is that the rich set of S-PLUS graphics functions are being leveraged to interpret the spline estimates found by a FUNFITS function. In this way, the FUN-FITS module is small, yet takes advantage of the many standard functions in S-PLUS.

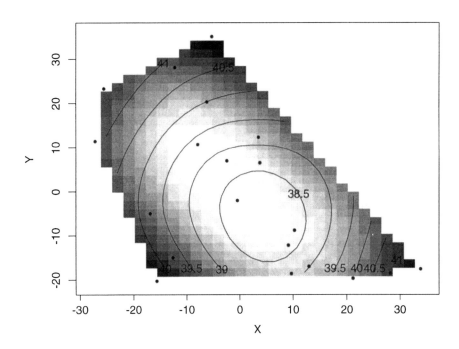

FIGURE 3. Estimated standard errors for mean ozone surface based on the GCV thin-plate spline (gray levels). Estimated mean levels for ozone are given by contour lines and the points are station locations.

3 A Basic Model for Regression

The suite of FUNFITS regression functions assume observations from the additive model,

$$Y_i = f(\boldsymbol{x}_i) + \epsilon_i \quad \text{for } 1 \leq i \leq n, \tag{A.1}$$

where Y_i is the observed response at the ith combination of design variables (or ith location), $\boldsymbol{x}_i \in \Re^d$, and f is the function of interest. The random components, $\{\epsilon_i\}$, usually associated with the measurement errors, are assumed be independent, zero-mean random variables with variances, $\{\sigma^2/W_i\}$.

One efficient strategy for representing f is as the sum of a low-order polynomial and a smooth function:

$$f(\boldsymbol{x}) = p(\boldsymbol{x}) + h(\boldsymbol{x}).$$

In the case of thin-plate splines (Wahba 1990; Green and Silverman 1994), the degree of p implies a specific roughness penalty on h based on integrated squared (partial) derivatives. Under the assumption that f is a realization of a spatial process, p can be identified as the "spatial trend" or "drift" and h is modeled as a mean zero Gaussian process (Cressie 1991). Both of these types of function estimates can be unified as ridge regression estimates (Nychka 1998). Here, the function is described as a linear combination of basis functions that either depend on the order of the spline or the covariance model, and the ridge regression constraint can be interpreted as a roughness penalty. This common structure in both problems was one motivation for the development of the FUNFITS numerical algorithms and their implementation in the S language.

Spline and spatial process models are sensitive to the dimension of x because they attempt to represent all possible interactions among the different variables. This feature is the well-known "curse of dimensionality" and can be tied to an exponential increase in the number of interactions as the dimension of x increases. Neural network regression Venables and Ripley 1994) is less sensitive to fitting high-dimensional functions. For regression problems, a useful form is a single, hidden-layer, feed-forward network

$$f(x) = \beta_0 + \sum_{k=1}^{M} \beta_k \phi(\mu_k + \gamma_k^T x), \qquad (A.2)$$

where

$$\phi(u) = \frac{e^u}{1 + e^u}$$

and β, μ, and γ_k, $1 \leq k \leq M$, are parameters to be estimated. In this form, the dimension reduction is achieved by only considering a small number of projections (M).

Another way to simplify the structure of f is by assuming that it is additive in the variables (Hastie and Tibshirani 1990):

$$f(x) = \sum_{k=1}^{d} f_k(x_k.) \qquad (A.3)$$

In FUNFITS, the full interactive model (A.1) can be estimated by the functions `tps` and `krig`, the neural net model (A.2) is fit by `nnreg`, and the additive model (A.3) by `addreg`. In addition there is a fast, one dimensional cubic spline function, `sreg`. It should be noted that within S-PLUS there exist excellent functions for fitting additive models and one-dimensional functions using splines (`gam`, `smooth.spline`). The main advantage of the FUNFITS version of these methods are efficiency and access to the all of the source code.

4 Thin-Plate Splines: `tps`

This section gives some background for the two surface-fitting functions, `tps` and `tpsreg`. Both of these fit thin-plate splines, the main difference being that `tps` is written essentially all in the S language and `tpsreg` has less options and executes a independent FORTRAN program (see Bates et. al. 1987) outside of S-PLUS.

Splines are usually defined implicitly as functions that solve a variational (minimization) problem. The estimator of f is the minimizer of the penalized sum of squares

$$S_\lambda(f) = \frac{1}{n}\sum_{i=1}^{n}(y_i - f(\boldsymbol{x}_i))^2 + \lambda J_m(f)$$

for $\lambda > 0$. The thin-plate spline is a generalization of the usual cubic smoothing spline with a "roughness" penalty function $J_m(f)$ of the form

$$J_m(f) = \int_{\Re^d}\sum \frac{m!}{\alpha_1!...\alpha_d!}\left(\frac{\partial^m f}{\partial x_1^{\alpha_1}...\partial x_d^{\alpha_d}}\right)^2 d\boldsymbol{x}$$

with the sum in the integrand being over all non-negative integer vectors, α, such that $\sum \alpha_1 + \cdots + \alpha_d = m$, and with $2m > d$. Note that for one dimension and $m = 2$, $J_2(f) = \int\left(f''(x)\right)^2 dx$, giving the standard cubic smoothing spline roughness penalty.

An important feature of a spline estimate is the part of the model that is unaffected by the roughness penalty. The set of functions where the roughness penalty is zero is termed the *null space*, and for J_m, it consists of all polynomials with degree less than or equal to $m - 1$.

Minimization of S_λ results in a function that tracks the data but is also constrained to be smooth. A more Bayesian viewpoint is to interpret S_λ as the negative of a posterior density for f given the data Y. The roughness penalty then is equivalent to a prior distribution for f as a realization from a smooth (almost m differentiable) Gaussian process. This is the basis for the connection between splines and spatial process estimates.

4.1 Determining the Smoothing Parameter

The discussion in the preceding section is given under the assumption that the smoothing parameter, λ, is known. In most applications, it is important to explore estimates of λ derived from the data. One standard way of determining λ is by generalized cross-validation (GCV). Although not obvious at this point, thin-plate splines are, in fact, linear functions of the observed dependent variable. Let $A(\lambda)$ denote the $n \times n$ smoothing or "hat" matrix that maps the Y vector into the spline predicted values:

$$(\hat{f}(\boldsymbol{x}_1),...,\hat{f}(\boldsymbol{x}_n))^T = A(\lambda)Y.$$

Given this representation, the trace of $A(\lambda)$ is interpreted as a measure of the number of effective parameters in the spline representation. The residuals are given by $(I - A(\lambda))Y$ and, thus, $n - \text{tr}A(\lambda)$ are the degrees of freedom associated with the residuals. The estimate of the smoothing parameter can be found by minimizing the GCV function

$$V(\lambda) = \frac{\frac{1}{n}Y^T(I - A(\lambda))^T W(I - A(\lambda))Y}{(1 - \text{tr}A(\lambda)/n)^2}.$$

One variation on this criterion is to replace the denominator by

$$[1 - (\mathcal{C}(\operatorname{tr} A(\lambda) - t) + t)/n]^2.$$

Here, t is the number of polynomial functions that span the null space of J_m and $\mathcal{C}$ is a cost parameter that can give more (or less) weight to the effective number of parameters beyond the base polynomial model (null space of the smoother). Once $\hat{\lambda}$ is found, σ^2 is estimated by

$$\hat{\sigma}^2 = \frac{Y^T(I - A(\hat{\lambda}))^T W(I - A(\hat{\lambda}))Y}{n - \operatorname{tr}(A(\hat{\lambda}))}.$$

and is analogous to the classical estimate of the residual variance from linear regression.

4.2 Approximate Splines for Large Data Sets

The formal solution of the thin-plate spline has as many basis functions as unique x vectors. Except in one dimension, this makes computation and storage difficult when the sample size exceeds several hundred. The basic problem is that the spline computation includes a singular value decomposition of a matrix with dimensions comparable to the number of basis functions. A solution to this problem in FUNFITS is to use a reduced set of basis functions, described by a sequence of "knot" locations, possibly different from the observed x values. Given below is a short example that goes through the steps of using a reduced basis.

The data set flame is an experiment to relate the intensity of light from ionization of lithium ions to characteristics of the ionizing flame. The dependent variables are the flow rate of the fuel and oxygen supplied to the burner used to ionize the sample and data consist of 89 points. Here is the fit to these data using a spline where the base model is a cubic polynomial:

```
flame.full.fit<- tps( flame$x, flame$y, m=4)
```

To reduce the number of basis functions, one could construct a grid of knots different from the observed values and smaller in number. The first step is to create the new grid of knots (a 6 × 6 grid in the example below) and then refit the spline with these knots included in the call to tps.

```
> g.list <- list(Fuel=seq(3,28,,6), O2=seq(15,39,,6))
> knots.flame <- make.surface.grid(g.list)
> flame.tps <- tps(flame$x,flame$y, knots=knots.flame, m=4)
```

Here, the grid list and make.surface.grid have been used to construct the two-dimensional grid. As an alternative, one could consider a random sample of the observed X values, knots.flame<-flame$x[sample(1:89,36),] or chose a subset based on a coverage design, knots.flame<- cover.design(

flame$x, 36). Of course, in either of these cases, the number of points, 36, is flexible and is based on choosing the number large enough to accurately represent the underlying function but small enough to reduce the computational burden.

4.3 Standard Errors

Deriving valid confidence intervals for nonparametric regression estimates is still an active area of research and it appears that a comprehensive solution will involve some form of smoothing that can vary at different locations. To get some idea of the precision of the estimate, in FUNFITS a standard error of the estimate is computed based on the assumption that the bias is zero. This is not correct and such confidence intervals may be misleading at points on the surface where there is fine structure such as sharp minima or maxima. However, empirical evidence suggests that such intervals provide a reasonable measure of the estimate's average precision if the smoothing parameter is chosen to minimize the mean squared prediction error. The standard error can be adjusted (inflated) by the factor $\sqrt{\text{tr}(A(\lambda))/\text{tr}(A(\lambda)^2)}$. This makes the average squared standard errors across the design points approximately equal to the average expected squared error. Some theoretical justification for this seemingly *ad hoc* fix is given in Nychka (1988). Operationally, one would follow the steps

```
> ozone.tps<- tps(ozone$x, ozone$y)
> ozone.se <- predict.surface.se(ozone.tps)
```

To adjust, the $z component of the surface object is multiplied by this inflation factor.

```
> inflate <- sqrt(ozone.tps$trace/ozone.tps$trA2)
> ozone.se$z <- ozone.se$z*inflate
> contour( ozone.se) # now plot adjusted standard errors
```

In some extreme cases, the actual coverage may drop from the 95% nominal level to 75%. However, in most situations such intervals are useful for assessing the order of magnitude of the estimation error.

5 Spatial Process Models: krig

The spatial process model used in this section assumes that $h(x)$ is a mean zero Gaussian process. Let

$$\text{Cov}(h(x), h(x')) = \rho k(x, x')$$

be a covariance function where k is a known function and ρ is an unknown multiplier. Letting $K_{ij} = k(x_i, x_j)$, the covariance matrix assumed for Y is

$\rho K + \sigma^2 W^{-1}$, and under the assumption that ρ and σ^2 are known, one can estimate $f(\boldsymbol{x})$ as the best linear, unbiased estimate for f based on Y (Cressie 1991; Welch et al. 1992; O'Connell and Wolfinger 1997). Recall that the basic model assumes that the measurement errors are uncorrelated with the ith observation having a variance given by σ^2/W_i. In spatial statistics, the variance of the errors is often referred to as the nugget effect and includes not only random variation due to measurement error but also fine-scale variability in the random surface in addition to the larger-scale covariance structure modeled by the covariance function.

To make the connection with ridge regression explicit, it is helpful to reparametrize the parameter ρ as $\rho = \lambda\sigma^2$. The estimation procedures in krig actually determine λ and σ^2 directly and then these estimates are used to derive the estimate for ρ.

Measures of uncertainty for $\hat{f}(\boldsymbol{x})$ are found by evaluating

$$E[(f(\boldsymbol{x}) - \hat{f}(\boldsymbol{x}))^2]$$

under the assumption that ρ and σ are known and that k is the correct covariance function. Because $\hat{f}(\boldsymbol{x})$ is a linear function of Y, it is a straightforward exercise to derive the prediction variance in terms of matrices that have been used to estimate the function. Of course, in practice, one must estimate the parameters of the covariance function, and in FUNFITS, the variance estimate is obtained by substituting these estimated values into the variance formula. The user should keep in mind that this is an approximation, underestimating the true variability in the estimate.

5.1 Specifying the Covariance Function

The strength of the krig function is its flexibility to handle very general covariance models. The key idea is that in the call to krig one must specify a covariance function written in the S language. This function should compute the cross-covariance between two sets of points. A basic form for the arguments is my.cov(x1,x2), where x1 and x2 are matrices giving two sets of locations and the returned result is the matrix of *cross*-covariances. Specifically, if x1 is $n_1 \times d$ and x2 is $n_2 \times d$, then the returned matrix will be $n_1 \times n_2$, where the i,j element of this matrix is the covariance of the field between locations in the ith row of x1 and the jth row of x2. Here is an example using the exponential covariance function

$$k(\boldsymbol{x},\boldsymbol{x}') = e^{-||\boldsymbol{x}-\boldsymbol{x}'||/\theta},$$

where the default value for the range parameter, θ, is set to 1.0:

```
> my.cov< function(x1,x2,theta=1.0){exp( -rdist(x1,x2)/theta)}
```

Here, rdist is a handy function that finds the Euclidean distance between all the points in x1 and those in x2 and returns the results in matrix form. For geographic coordinates, `rdist.earth` gives great circle distances for locations reported in longitude and latitude, and because we will use such a function later, we also give its definition in S.

```
> exp.earth.cov<- function(x1,x2,theta=1.0)
  {exp( -rdist.earth(x1,x2)/theta)}
```

A function that has a few more options is the FUNFITS version `exp.cov` and this could be used as a template to make other covariance functions. At a more complex level is `EXP.cov` that makes use of FORTRAN routines to multiply the covariance function with a vector. Although more difficult to read, these modifications decrease the amount of storage required for evaluating the kriging or thin-plate spline estimate at a large number of points.

Most covariance functions also require a range parameter, e.g., θ in the case of the exponential. The parameters that appear in a nonlinear fashion in k are *not* estimated by the `krig` function and must be specified. The parameter values can be given in two ways: either by modifying the covariance function directly or changing the default value for a parameter when `krig` is called. Suppose that one wanted to use an exponential function with $\theta = 50$. One could either modify `my.cov` so that the default was `theta= 50` or in the call to `krig` include the `theta` argument:

```
> ozone.krig <- krig(ozone$x, ozone$y, my.cov, theta=50)
```

In general, any extra arguments given to `krig` are assumed to be parameters for the covariance function. The function `krig` makes a copy of the covariance function, substitutes these new defaults for the parameters, and includes this modified function as part of the output object in the `cov.function` component. For example, looking at the `krig` object created above,

```
> ozone.krig$cov.function
function(x1, x2, theta = 50)
{
        exp( - rdist(x1, x2)/theta)
}
```

This is identical to the orignal version of the function, except the default value for `theta` has been changed! In this way, subsequent computation based on the krig object will use the correct parameter value. At a more abstract level one is not limited to passing just single values as parameters to a covariance function. For more complicated covariance models that require many pieces of information one can pass objects or lists to specify the form. See as an example the FUNFITS function `kernel.cov`.

There are well-established methodologies for estimating the parameters of the covariance function and perhaps the most straightforward is nonlinear least squares fitting of the variogram. FUNFITS has a variogram function, `vgram` and many estimation methods are based on estimating parameters from the variogram using nonlinear least squares. Because S already provides support for nonlinear least squares fitting, it is not neccesary to develop specialized variogram-fitting routintes within FUNFITS. The example in the next section fits a variogram in this way using the S-PLUS function nls.

In contrast to the nonlinear parameters of the covariance function, the default for krig is to estimate the parameter λ (σ^2/ρ) based on cross-validation.[2] This difference is based on the fact that the qualitative aspects of the estimated surface are usually much more sensitive to the value of the smoothing parameter, λ, than the others. Also, this parameter appears in a linear fashion in the matrix expressions for the estimate, and so it is efficient to evaluate the estimate at different choices of this parameter. Parameters such as the range do not have a similar property, and the matrix decompositions for the estimates must be recalculated for different values of these parameters.

5.2 Some Examples of Spatial Process Estimates

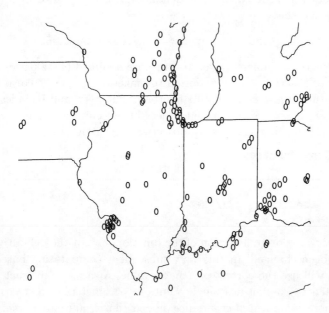

FIGURE 4. Locations of ozone monitoring stations for the midwestern United States.

---

[2]In the next version, we hope to add a maximum likelihood method for determining λ.

The data set `ozone2` contains the daily, 8-hour average ozone measurements during the summer of 1987 for approximately 150 locations in the Midwest (Figure 4). The component y in this data set is a matrix of ozone values with rows indexing the different days and columns being the different stations. The locations can be plotted by

```
> US( xlim= range( ozone2$lon.lat[,1]),
    ylim= range( ozone2$lon.lat[,2]))
> points( ozone2$lon.lat, pch="o")
```

Most of the work in this example is identifying a reasonable covariance function for these data. Here is an example of computing and plotting the variogram for day 16:

```
> y16<- c(ozone2$y[16,]
> vgram.day16<- vgram( ozone2$lon.lat, y16, lon.lat=T)
> plot( vgram.day16, xlim=c(0,150))
```

The variogram in this raw form is not very interpretable and one could use different smoothing methods such as *loess* or spline smoothing to smooth these points.

As an alternative, one could look at the correlogram to get an idea of the general distance scale for the correlations. Under the assumption that ozone over this period has a consistent spatial covariance from day to day, one can consider the different days as replicated points and estimate the correlations directly. The following S code finds the pairwise correlations and plots them against distance:

```
> ozone.cor<- COR( ozone2$y)
> upper<- col(ozone.cor$cor)>row(ozone.cor$cor)
> cgram.oz<- list( d=rdist.earth( ozone2$lon.lat)[upper],
                  cor= ozone.cor$cor[upper])
> plot( cgram.oz$d, cgram.oz$cor, pch=".")
```

Here, we have used the FUNFITS function COR because it can handle missing values (in S-PLUS, NAs). This is important because the missing value patterns are irregular and reducing down to a complete data set would exclude many of the days. This function returns a list that includes the correlation matrix ($cor) the sample means ($means), and sample standard deviations ($sd). One disadvantage is that COR is slow because it loops through all pairs of stations. Next, one can fit an exponential correlation function to these sample correlations using nonlinear least squares:

```
> cgram.fit<- nls( cor ~ alpha* exp( -d/theta),
        cgram.oz, start= list( alpha=.95, theta=200))

> summary( cgram.fit)
```

Formula: cor ~ alpha * exp(- d/theta)

Parameters:
```
          Value    Std. Error  t value
alpha   0.931247  0.00366247  254.268
theta 343.118000  2.37364000  144.554
```

Residual standard error: 0.120286 on 11626 degrees of freedom

Correlation of Parameter Estimates:
```
        alpha
theta  -0.838
```

Based on this value for the range, one can now remove the missing values for day 16 and use the krig function to fit a surface:

```
> good<-!is.na(y16)
> ozone.krig<- krig( ozone2$lon.lat[good,], y16[good],
         exp.earth.cov, theta=343.1183)
```

In this case, the nugget variance and the linear parameters of the covariance are found by GCV. Here is a summary of the results:

```
> summary(ozone.krig)
Call:
krig(x = ozone2$lon.lat[good, ], Y = y16[good],
  cov.function = exp.earth.cov,theta = 343.1183)
```

```
Number of Observations:                         147
Degree of polynomial null space ( base model):   1
Number of parameters in the null space           3
Effective degrees of freedom:                  108.1
Residual degrees of freedom:                    38.9
MLE sigma                                       6.306
GCV est. sigma                                  6.647
MLE rho                                         2648
Scale used for covariance (rho)                 2648
Scale used for nugget (sigma^2)                39.77
lambda (sigma2/rho)                             0.01502
Cost in GCV                                     1
GCV Minimum                                    166.8
Residuals:
   min   1st Q  median  3rd Q    max
  -16.4 -1.697  0.0953  1.685  12.15
```

For this particular day, the ozone surface has a high degree of variability, more so than a typical day in this period, and this is reflected by the large value of the rho parameter in the summary. Alternatively, one could specify sigma2 and rho explicitly in the call, and thus fix them at specific values. More is said about this below.

The next set of S calls evaluate the fitted surface and prediction standard errors on a grid of points. These evaluations are then converted to the format for surface plotting and some contour plots are drawn (see Figure 5):

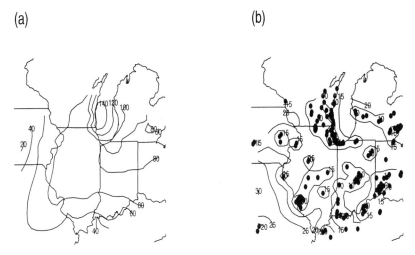

FIGURE 5. Fitted surface and standard errors of the Midwest ozone network using an isotropic covariance function.

```
> set.panel(2,1)

> out.p<- predict.surface( ozone.krig)
> US( xlim= range( ozone2$lon.lat[,1]),
      ylim= range( ozone2$lon.lat[,2]))
> contour( out.p, add=T)

> out2.p<-predict.surface.se( ozone.krig)
> US( xlim= range( ozone2$lon.lat[,1]),
      ylim= range( ozone2$lon.lat[,2]))
> contour( out2.p,add=T)
> points( ozone2$lon.lat)
```

The last part of this section illustrates how to extend this analysis to a covariance function that is not stationary. One simple departure from nonstationarity is to consider a covariance that has different marginal variances but isotropic correlations. Specifically, let $Y(\boldsymbol{x})$ denote the value of ozone at location $\boldsymbol{x}$ and assume that $EY(\boldsymbol{x}) = \mu(\boldsymbol{x})$ and $Var(Y(\boldsymbol{x})) = \sigma(\boldsymbol{x})^2$. Thus, μ and σ are the marginal mean and standard deviation, respectively, for the spatial process. Now, consider the standardized process

$$Z(\boldsymbol{x}) = \frac{Y(\boldsymbol{x}) - \mu(\boldsymbol{x})}{\sigma(\boldsymbol{x})}.$$

The key assumption is that this new process is isotropic. In Particular, assume that $Z(\boldsymbol{x}_k) = f(\boldsymbol{x}_k) + e_k$, where f has an exponential covariance kernel, the same one that has been used in the preceding example. The estimates for the ozone field under this setup are easy. If μ and σ are known, one standardizes the observed data, computes the spatial process, estimate for the standardized, isotropic process and applies the inverse transform to obtain predictions for the original process; that is, $\hat{Y}(\boldsymbol{x}) = \hat{Z}(\boldsymbol{x})\sigma(\boldsymbol{x}) + \mu(\boldsymbol{x})$. The standard errors for $\hat{Y}(\boldsymbol{x})$ are those derived for Z multiplied by $\sigma(x)$. [3]

The function krig has been structured to make such correlation models with which it is easy to work. As a first step, we can estimate the marginal mean and standard deviation surfaces using the different days in the ozone record as replicates. The means and standard deviations have been computed previously using COR and so we will just use those results here:

```
> mean.tps <- tps( ozone2$lon.lat, ozone.cor$mean,
      return.matrices=F)
> sd.tps<- tps( ozone2$lon.lat, ozone.cor$sd,
      return.matrices=F)
```

The return.matrices switch has been set to false so that the large matrices needed to find standard errors for the spline are not included in the output object. We do not need to calculate standard errors, and using this smaller object is more efficient. In this example, just a thin-plate spline is used to represent these two surfaces. However, one could also use spatial process estimates to find these functions. This would be useful if one also wanted to include the uncertainty in estimating the mean function in the final prediction standard errors. The reader is referred to the krig help file for more details on this approach.

The next step is to call krig with a covariance function that uses the two tps objects to evaluate the marginal means and standard deviation surfaces. We can use the same exponential covariance function used from the previous example because the parameters were estimated from the correlations.

---

[3] In this example, we are avoiding the more difficult problem of adjusting the prediction standard errors for the fact that μ and σ are estimated and not known precisely. The help file on predict.se.krig gives some extensions to account for this problem.

To use krig to fit the surface, one has two options. The first is to keep all of the covariance parameters fixed using the values estimated from the correlogram. Here is how to specify and fit this model:

```
> fit.fixed<- krig( ozone2$lon.lat[good,], y16[good],
    rho=.93, sigma2= (1-.93),
    cov.function=exp.earth.cov,theta=343,
    sd.obj=sd.tps, mean.obj= mean.tps,
    m=1)
```

Note the addition of the two thin-plate spline objects that describe the mean and standard deviation surfaces. Finally, because this model already includes a mean function, μ, only a constant term is used for the polynomial part of the model ($m = 1$).

The second approach is to estimate the parameters ρ and σ from the data but to keep the rest of the covariance fixed. The S code below is for finding both parameters using GCV:

```
> fit.GCV<- krig( ozone2$lon.lat[good,], y16[good],
    cov.function=exp.earth.cov,theta=343,
    sd.obj=sd.tps, mean.obj= mean.tps, m=1)
```

This gives a different fit to the data, with the estimate for ρ being substantially larger, and, similar to the results from before, reflects the fact that for this given day the ozone surface has more variability than on average. To compare the two different models, we can make a panel of plots given in Figure 6.

```
set.panel(2,2)
out1.p<-predict.surface(fit.fixed)
out2.p<-predict.surface(fit.GCV)
out3.p <- predict.surface.se(fit.fixed)
out4.p <- predict.surface.se(fit.GCV)
surface(out1.p, type="c")
surface(out2.p, type="c", )
surface(out3.p, graphics.reset=F, type="c")
points(ozone2$lon.lat)
surface(out4.p, graphics.reset=F, type="c")
points(ozone2$lon.lat)
```

Here, the difference in ρ between the two models has increased the standard errors associated with the GCV fit by about a factor of three. This example also helps to illustrate the use of some of the FUNFITS functions for plotting surfaces and adding information to surface plots.

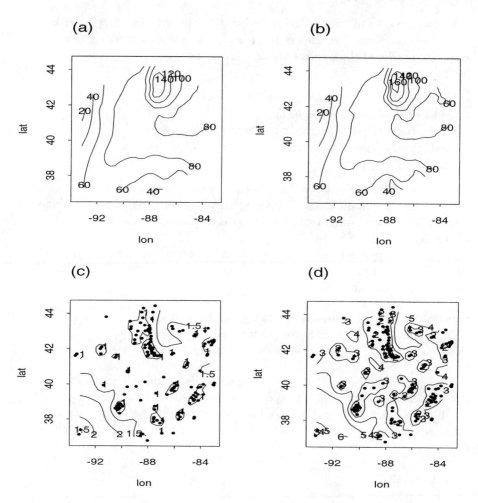

FIGURE 6. Fitted surfaces and standard errors of the Midwest ozone network using an isotropic correlation function. Plot (a) is the predicted ozone surface for day 16 using parameters for the covariance, σ, and ρ estimated from the 89 days of the summer. Plot (c) gives the corresponding prediction standard errors. Plot (b) gives the predicted ozone surface where the parameters have been estimated for this particular day using GCV and plot (d) gives the corresponding prediction standard errors.

Acknowledgments

This work was supported in part by the National Science Foundation DMS-9217866-02 and Becton Dickinson Research Center.

References

Bates, D.M., Lindstrom, M.J., Wahba, G. and Yandell, B.S. (1987). GCV-PACK - Routines for generalized cross validation. *Communications in Statistics - Simulation and Computation* **16**, 263–297.

Cressie, N.A.C. (1991). *Statistics for Spatial Data.* Wiley, New York.

Green, P.J. and Silverman, B.W. (1994). *Nonparametric Regression and Generalized Linear Models.* Chapman & Hall, London.

Hastie, T. and Tibshirani, R. (1990). *Generalized Additive Models.* Chapman & Hall, New York.

Nychka, D. (1988). Bayesian "Confidence" intervals for smoothing splines. *Journal of the American Statistical Association* **83**, 1134–1143.

Nychka, D., Bailey, B., Ellner, S., Haaland, P. and O'Connell, M. (1996). FUNFITS data analysis and statistical tools for estimating functions. Technical Report No. 2289. North Carolina Institute of Statistics Mimeo Series, Raleigh, NC.

Nychka, D. and Boos, D. (1997). S-LAB: Instructional labs on data analysis and graphics using S. Technical Report No. 2292. North Carolina Institute of Statistics Mimeo Series, Raleigh, NC.

Nychka, D. (1998). Spatial process estimates as smoothers. *Smoothing and Regression. Approaches, Computation and Application.* ed. M. G. Schimek, Wiley, New York.

O'Connell, M. and Wolfinger, R. (1997). Spatial regression, response surfaces and process optimization. *Journal of Computational and Graphical Statistics* (in revision).

Venables, W.N. and Ripley, B.D. (1994). *Modern Applied Statistics with S-Plus.* Springer-Verlag, New York.

Wahba, G. (1990). *Spline Models for Observational Data.* Society for Industrial and Applied Mathematics, Philadelphia.

Welch, W.J., Buck, R.J., Sacks, J., Wynn, H.P., Mitchell, T.J. and Morris, M.D. (1992). Screening, predicting and computer experiments. *Technometrics* **34**, 15–25.

Appendix B: DI, A Design Interface for Constructing and Analyzing Spatial Designs

Nancy Saltzman
National Institute of Statistical Sciences

Douglas Nychka
North Carolina State University

This appendix describes the DI (Design Interface) package for interactive graphical spatial design, which grew out of the work presented in Chapter 4. It contains some portions of the DI users manual which the more interested reader may obtain, along with the DI package, from the DI homepage http://www.ncsu.stat.ncsu/~nychka/di or through the NISS homepage http://www.niss.org/

1 Introduction

The DI package of S-PLUS programs was motivated by the need for a flexible graphical tool to evaluate performance of spatial designs. Its original application was in analyzing networks of environmental monitoring stations. Although several researchers have created algorithms to calculate optimal designs under different criteria, there is a practical need to determine a given design's robustness to other measures of performance and to understand the effect of adding and deleting points from an "optimal" set of locations. Also, in many applications, spatial networks of monitors are already in place and it is of interest to analyze the effect of thinning such networks or moving locations. DI is intended to facilitate construction and comparison of spatial designs by visualization of the designs themselves and of the statistics that measure their performance. The statistical theory underlying DI has been presented in Chapter 4. In particular, the material on spatial models in Chapter 4, Section 3 is the statistical basis for the DI computations.

DI operates with a menu interface (as shown in Figures 1 and 2) that is accessible to all potential users, including those with no programming or S-PLUS experience. However, DI can also be easily customized to include menu buttons for any user-programmed operations. The basis for the DI structure is the network object `nw` and network catalog object `nw.catalog`, data sets which include all information about a network design and allow use of S-PLUS object-oriented features.

2 An Example

Figures 1 and 2 show an analysis of the Chicago ozone monitoring station network in which the *leaps* algorithm (Chapter 4, Section 4.2) is used to thin the original network and resulting designs are compared. In Figure 1, DI was started from the hpterm window (the HP equivalent of an xterm window) in the lower right, which continues to display brief instructions throughout the DI session. DI's Main Menu, upper left, contains buttons for all the DI functions and any additional user functions. The Main Window in the upper right is the working window for DI in which stations are added and deleted. The `Plot` function has been selected from the Main Menu and is currently active. The Plot window shows the network and its associated prediction variance surface. Components of the Plot display are controlled by the Replot menu. Figure 2 shows three results of the leaps thinning algorithm applied to the network of Figure 1. The new networks, sizes 20, 15, and 12 stations, and their prediction variances, are displayed in the Plot windows. The Plot Summary window in the upper right displays mean and maximum prediction variance for the different sized networks. Note that the 15-station network performs very similarly to the 20-station network.

3 How DI Works

3.1 Network Objects

The central concept of DI is the network object `nw` used to bundle together all the information that is required to evaluate and visualize a design. It includes components for network station locations, model covariance, evaluation grid, and summary statistics. The network catalog object `nw.catalog` keeps several network objects together. DI uses the object oriented tools in S-PLUS to operate on network and network catalog objects. For example, the command `print(square.nw)` is equivalent to `print.nw(square.nw)` because `square.nw` is of class `nw` and DI has a special-purpose print function for these objects. Thus, a network object does not print as a list. The functions `summary` and `plot` operate in a similar manner. The functions `print` and `summary` have also been extended to network catalog objects.

3.2 The Design Editor

The design editing function, `edit.nw` or `edit`, initiates the Motif dialog box interface (Figures 1 and 2) through which user interaction with DI takes place. The Main Menu has actions to modify, summarize, plot, and save designs, with secondary menus generated as appropriate. The editor takes a network or network catalog object as input and displays one of the network objects. The Main Menu is then used to operate on this design object, applying a function

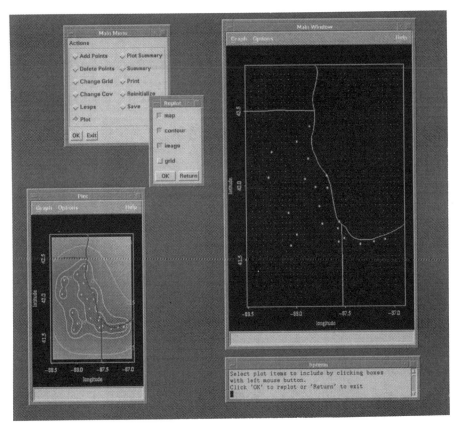

FIGURE 1. DI session analyzing the Chicago ozone network.

to it or modifying it to create a new network. When a modified design is saved, it is added to the current working network catalog. Finally, when the editor function is quit, it returns a network catalog object.

3.3 User Modifications

DI has been written so that it is easy to add new actions to the Main Menu. Each button corresponds to a few lines of S code. The main loop in the editor identifies the selected button and evaluates the S code that corresponds to that button. The list linking button names (the *items* in the Main Menu) to S code that performs the correct operations is in the data set di.items. Thus, adding new functionality to DI is simply a matter of editing the di.items file within S to include a button to access the new code.

An initial section of the di.items data set is given here for illustrative purposes. There are two working data sets that are used within the editor. nw is the name of the current network object. If any changes are made to nw, then the component nw$id is set to zero so the Save function will recognize this

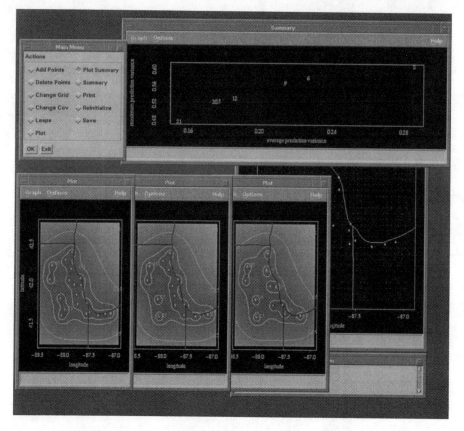

FIGURE 2. Results of thinning the Chicago network using the leaps algorithm.

design as a new one and save it. design.history is a network catalog object that contains the initial network object and any others that are saved in the course of using the editor.

```
"di.items"<-
expression("Add Points" = {
nw <- di.add(nw)
nw$id <- 0
}
, "Delete Points" = {
nw <- di.del(nw)
nw$id <- 0
}
, "Change Grid" = {
nw <- di.chgrid(nw)
}
```

```
, "Change Cov" = {
temp <- list(n = nw, history = history.design)
temp <- di.chcov(temp)
nw <- temp$n
history.design <- temp$history
nw$id <- 0
}
    .
    .
    .
```

Appendix C: Workshops Sponsored Through the EPA/NISS Cooperative Agreement

Statistical Methods for Combining Environmental Information (1993)

Organizers: Jerome Sacks, Director NISS; Lawrence Cox, Senior Statistician, U.S. EPA.
Ecological Assessments and the Need to Combine Them. R. Linthurst, EPA
Combining Sample-Based and Observational Data for Ecological Assessment. T. Olsen, EPA
Combining Epidemiological Studies for Human Health Risk Assessment: Passive Smoking and Dioxin. S. Bayard, EPA
Synthesis of Environmental Evidence: Nitrogen Dioxide Epidemiology Studies. D. Kotchmar, EPA
Combining the Three-Cities Lead Abatement Studies. A. Marcus, EPA
Beyond Benchmark: A Bayesian Application to Dose-Response Analysis of Non-cancer Toxicity. A. Jarabek, EPA, and V. Hasselblad, Duke
Developing an Acute Inhalation Assessment Method. D. Guth, EPA
Combining Independent F-Tests with Environmental Application. B. Sinha, UMBC
Mapping Procedures Using Multiple Data Sets. N. K. Nagaraj, UMBC, and R. Shafer, EPA
International Data Conformance. B. Bargmeyer, EPA
Developing Environmental Indicators and Indexes. E. Hyatt, EPA, and D. Hoag, Colorado State

Spatial Sampling for the Environment (1994)

Organizers: Larry Cox, U.S. EPA; G.P. Patil, Penn State; Jerome Sacks, NISS.
Spatial Sampling Designs—Description, Examples, Questions. Noel Cressie, Iowa State
Spatial Designs—Methodology Update, Evaluation, Future Issues. Don Ylvisaker, UCLA
Additional Comments on Spatial Design. Nhu Le, UBC, and Keh-Shin Lii, UC-Riverside
National Resources Inventory - Description, Future Issues. Wayne Fuller and Jay Breidt, Iowa State
Spatial Sampling for Surface Soil Remediation. Evan Englund, EPA
Spatial Problems in Waste Site Characterization. John Warren, EPA
Additional Comments on Site Remediation and Characterization. Steve Thompson, Penn State and Max Morris, ORNL

Statistical Issues in Mechanistic Modeling for Risk Assessment (1995)

Organizers: Jerome Sacks, NISS; Larry Cox, U.S. EPA; Chris Portier, NIEHS.
Mechanistic Modelling of DCM. Mel Anderson, ICF Kaiser-KS Crump Division
Biological Mechanisms and Use of DCM Models. Ray Yang, Colorado State
DCM Model Uncertainty and Validation. Harvey Clewell, ICF Kaiser
Biological Mechanisms. Linda Birnbaum, EPA-NHEERL
TCDD Mechanistic Modelling. Michael Kohn, NIEHS-LQCB
Uncertainty Assessment and Validation for TCDD Models. Chris Portier, NIEHS-LCQB
Tox-Epi Model Linkage: Examples. Lutz Edler, German Cancer Research Center
Cross-species Extrapolation of Mechanistic Models. Lorenz Rhomberg, Harvard Center for Risk Analsis
Developing Toxicological Equivalency Factors. Michael DeVito, EPA-NHEERL
Sensitivity and Uncertainty Analysis in Mechanistic Models. Marina Evans, EPA-NHEERL
The Biokinetic Lead Uptake Model. Robert Elias and Allan Marcus, EPA-NCEA
Validating the Biokinetic Lead Uptake Model. Dennis Cox, Rice University

Statistical Issues in Setting Air-Quality Standards (1996)

Organizers: Jerome Sacks, NISS; Larry Cox, U.S. EPA.
Overview: Air Quality Standards And Spatial Averaging. Karen Martin, U.S. EPA and Harvey Richmond, U.S. EPA
Statistical Issues in Exceedance Criteria. Ross Leadbetter, UNC-Chapel Hill
Preliminary Results of Derived PM2.5 Spatial Averages. Terence Fitz-Simons, U.S. EPA
Framework for Measuring Exposure: A Space-Time Ozone Surface for Houston. Katherine Ensor, Rice University
Forecasting Different Measures of Exposure. Paul Speckman, University of Missouri
Air Quality and Emergency Room Visits for Asthma in Atlanta. David MacIntosh, University of Georgia
Measuring Ozone Exposure for Vegetation. E. Henry Lee, Dynamic International, Inc.
Particulates and Mortality. Richard Smith, UNC-Chapel Hill
Monitoring Issues Associated with Spatial Averaging. Neil Frank, U.S. EPA
Impact of Exposure Measures on Network Design. Doug Nychka, NCSU

Appendix D: Participating Scientists in the Cooperative Agreement

Markus Abt National Institute of Statistical Sciences
Roger Berger North Carolina State University
Raymond Carroll Texas A & M University
Shao-Hang Chiu U.S. EPA
Dennis Cox Rice University
William Cox U.S. EPA
John Creason U.S. EPA
J. Michael Davis U.S. EPA
William Du Mouchel AT & T
Evan Englund U.S. EPA
Kathy Ensor Rice University
Terence Fitz-Simons U.S. EPA
George Flatman U.S. EPA
Ralph Folsom Research Triangle Institute
Gary Foureman U.S. EPA
Feng Gao National Institute of Statistical Sciences, Battelle Northwest
Steve Garren National Institute of Environmental Health Sciences
Vincent Granville National Institute of Statistical Sciences
Dave Guinnup U.S. EPA
Dan Guth U.S. EPA
David Higdon Duke University
James Hilden-Minton National Institute of Statistical Sciences
Marge Holland U.S. EPA
Eric Hyatt U.S. EPA
Jeff Jonkman North Carolina State University
Dennis Kotchmar U.S. EPA
Ross Leadbetter University of North Carolina at Chapel Hill
Teressa Leavens University of North Carolina at Chapel Hill
Sharon LeDuc U.S. EPA
Allan Marcus U.S. EPA
Nancy McMillan National Institute of Statistical Sciences
John Monahan North Carolina State University
Lisa Moore Duke University
Max Morris Oak Ridge National Laboratory
William Nelson U.S. EPA
Jan Novella Institute for Transportation Research & Education
Gary Oehlert University of Minnesota
Kenneth Poole Research Triangle Institute
Norm Possiel U.S. EPA OAQPS

John Rawlings North Carolina State University
Steve Rembish Texas Natural Resource Conservation Commission
Andy Royle North Carolina State University
Ken Schere National Oceanic and Atmospheric Administration/U.S. EPA
Chon Shoaf U.S. EPA
Jack Shreffler U.S. EPA
Douglas Simpson University of Illinois
Laura Steinberg National Institute of Statistical Sciences
Patricia Styer National Institute of Statistical Sciences
David Svendsgaard U.S. EPA
Lance Waller University of Minnesota
William Welch University of Waterloo
Steve Williams University of California-Los Angeles
Robert Wolpert Duke University
Minge Xie National Institute of Statistical Sciences
Qing Yang North Carolina State University
Donald Ylvisaker University of California-Los Angeles
Jun Zhai North Carolina State University, U.S. EPA
Haibo Zhou National Institute of Environmental Health Sciences

Index

ED_{100q}, see Effective dose
ED_{100q} line, 129
CatReg package, 134
Loess smoother, 11, 39, 96, 99, 110
P-values
    combining, 145, 146
        Fisher's method, 145, 152
        inclusion probabilities, 154
        poststratification, 154
lasso procedure, see Subset selection
leaps procedure, see Subset selection

DI package
    Web Site, 181

Abbott's formula, 133
Air-quality monitoring, 51
Air-quality monitoring network, 2
ARMA(1,1) model, 85
Atmospheric deposition, 2, 77
Atmospheric particulates, see Particulate matter

B-spline, see Spline function
Backward selection, see Variable selection
Basis functions, 160
Bayesian analysis
    hierarchical models, 150
    shrinkage, 150
Benchmark dose, 122
Benthic index, 151

Case studies, see Studies

Censoring, interval, 125, 127
Clean Air Act
    Great Britain, 91
    United States, 6, 77, 91, 121
Cluster analysis, 37–38
    k-means, 37, 42
Clustering algorithm
    average linkage, 37
    complete linkage, 37
    single linkage, 37
Combining environmental information, 3, 144–146
Component scores matrix, 31
Correlated observations
    and overdispersion, 131
Covariance function, 169, 170
    isotropic, 57, 176
    nonstationary, 59
    range parameter, 59, 74, 170, 171
    semiparametric, 59
Covariance-filling design, 71
Cross-validation
    generalized, 167
    predictive, 18, 23
Curve fitting
    via FUNFITS, 160

Data matrix
    for singular value decomposition, 30
    mode of variation in, 31
Design of experiments
    A-optimality, 53
    G-optimality, 53
    spatial, 53, 74, 181

Deviance, 128
DI package, 181–185
Distributions
    double exponential family, 156
    generalized extreme value, 19
    generalized Pareto, 20
Dose-response model, *see Exposure-response*

Effect sizes
    combining, 145
Effective dose (ED_{100q}), 128–130
EMAP, *see Environmental Protection Agency*
Environmental assessment, 1, 144, 150
    combining information, 150
Environmental epidemiology, 145
Environmental monitoring, 1, 51
    combining information, 153
    network design, 53, 181
Environmental Protection Agency (U.S.), 1
    acute toxicity database, 122, 124
    Aerometric Information Retrieval System, 8
    Environmental Monitoring and Assessment Program (EMAP), 144
    EPA/NISS Workshops, 143, 187
Environmental tobacco smoke (ETS), *see Passive smoking*
Environmental toxicology, 3
Error sum of squares, 10
Euclidean distance, 58
    via S-PLUS `rdist` function, 170
Exceedance modeling, 19–22
Exceedance over threshold, 20
Experiment design, *see Design of experiments*

Exposure measurements
    concentration, 122
    duration, 122
Exposure-response
    combining information, 122
    marginal modeling, 131–132
        and generalized estimating equations, 131
    severity scoring, 125
    worst-case analysis, 133, 135

FUNFITS package, 39, 159–176
    Web Site, 160

Gamma regression model, *see Regression model*
    for daily sulfate concentration, 81
Gaussian additive spatial process, 16
Generalized additive model (GAM), 39, 44, 108
Generalized cross-validation (GCV), *see Cross-validation*
Generalized estimating equations (GEEs), 131

Haber's law, 126–127
    linearized, 127
Hierarchical model, 149
Human equivalent concentrations (HECs), 138

Influence regimes
    geographic, 48
Inverse variance weighting, 149, 154
Isotropy, 57

Jackknifing, 14, 82

Kriging, 16, 60, 151, 153

Lagged variable, 99

Index

Linear regression, *see Regression model*
Link function
   in gamma regression model, 81
   in ordinal regression model, 127
Logit model, 20

MAP3S monitoring network, 78
Marginal modeling, *see Exposure-response*
Markov dependence, 23
Matrix
   component scores, 31
Median polish, 11, 56
Meta-analysis, 145
Meteorological variables, 8–10, 40
   atmospheric pressure, 9
   ceiling height, 9
   cloud cover, 9
   dew-point temperature, 9, 94, 99
   mixing height, 9
   opaque cloud cover, 9
   pressure, 112
   relative humidity, 9, 18
   specific humidity, 9, 94, 99, 109, 112
   temperature, 9, 18, 81, 109, 112
      lagged, 116
      maximum, 94, 99, 109
      minimum, 94, 99
   total precipitation, 81
   visibility, 9
   wind, 9, 18, 81
Monitoring network
   augmenting, 71–86
   thinning, 62, 83
      NADP/NTN, 86–87

NAMS/SLAMS monitoring network, 44, 55–56, 69
National Ambient Air Quality Standards, 6
National Institute of Statistical Sciences, 1
   EPA/NISS Workshops, 143, 187
   Web Site, 4, 160, 181
National Resources Inventory (U.S.), 153
National Weather Service (U.S.), 152
   Eta model, 49
Network design, 2, 53–54, 74, 85, 181
Neural network regression, *see Regression model*
No-observed-adverse-effect level (NOAEL), 122
Nonlinear regression model, 10
Nonparametric regression, 38, 108
   and spline functions, 38, 109
   and *loess* smoothing, 39

Odds ratios, 146
   combining, 146, 149
Overdispersion, 131, 156
Ozone, 5
   adjusted trend analysis, 18
   diurnal behavior, 29
      model for, 34
   diurnal profile from hourly measurements, 33
   monitoring network, 51
   network typical value, 11, 17
   spatial distribution, 52, 54
   standards for ambient concentrations, 6
   surface-level, 2
   temporal seasonality, 46
Ozone cluster
   anticyclonic, 44
Ozone season, 8, 11

P-values
   Bonferroni adjustment, 116

Particulate matter
    and confounding effects, 93
    and mortality, 91–92
    health effects, 119
    PM_{10}, 93–95
    $PM_{2.5}$, 94
    total suspended particulates, 93
Particulate measurements, 2
Passive smoking, 146
Perchloroethylene, *see Studies*
Physiologically based pharmacokinetic (PBPK) models, 138
Piecewise linear model
    for mortality, 99, 101
    for relative risk, 105
PM_{10}, *see Particulate matter*
Poisson process, 22
Poisson regression, *see Regression model*
Prediction variance, 60
Principal components
    rotated, 45
Principal components analysis, 31, 45
    Varimax criterion, 45
Probability samples (P-samples), *see Sampling*
Proportional odds ratio, 127
Pseudo-F statistic, 41

Random field, 57–59
    nonstationary model, 59
Ranked set sampling, *see Sampling*
Regional Acid Deposition Model (RADM), 84–85
Regional Oxidant Model (ROM), 53, 55, 57
Regression model
    additive, 165
    categorical, 126
        ordinal levels, 126

gamma, 80
    for daily sulfate concentration, 80
    for weekly sulfate concentration, 82
hierarchical, 149
linear, 10, 95
logistic, 127
neural network, 166
P-samples
    augmenting, 155
Poisson, 96
    seasonality adjustment, 112
stratified, 129–130
Relative risk, 103, 146
    adjusted, 148, 150
    combining, 149
Risk assessment
    exposure-response, 123
    quantitative, 121–123

S-PLUS, 64
    CatReg package, 134
    leaps function, 63, 64
    `addreg` function, 40
    `loess` function, 97
    `bs` function, 98
    `gam` function, 109, 166
    `glm` function, 81, 96
    `loess` function, 105, 109, 160
    `nls` function, 172
    `smooth.spline` function, 166
    DI package, 65, 181–182
    FUNFITS functions, 159–160, 164
    S-Lab introduction, 160
Sampling
    adaptive, 154
    dual-frame estimation, 154
    NP-samples, 153
        combining, 155–156
    P-samples, 153
        combining, 153–154
    ranked set, 155

Index 195

Sampling *continued*
    spatial, 151, 153
SAS™, 128
    Proc Logistic, 130
Scree test, 41
Seasonal interaction, 116
Semiparametric modeling, 16, 39
Severity scoring, *see*
    Exposure-response
Singular value decomposition
    (SVD), 30–31, 45
Snow water equivalent, 152
Space-filling design, 68–69, 159
    distance metric, 68
    minimax, 69
Spatial analysis
    Kaiser rotation, 46
Spatial prediction, 16, 57, 60
Spatial process models, 169–172
    curse of dimensionality, 166
    GaSP, 16
Spatiotemporal model, 83
Spline function, 38, 166
    B-spline
        approximation for relative
            risk, 105
        equally spaced knots, 96–98
    curse of dimensionality, 166
    knots, 38, 98
    roughness penalty, 39, 167
    smoothing, 38, 167
    smoothing parameter, 167
    thin plate, 161, 165–169
Studies
    Birmingham
        particulate–mortality, 94,
            102–104
    Chesapeake Bay benthos, 150
    Chicago area ozone, 7
    Chicago particulate–mortality,
        112
        reanalysis, 114
    Great Lakes ozone network, 71
    Gulf Coast states ozone, 46
    Houston area ozone, 7
    Houston hourly ozone, 31
    Houston urban ozone, 41
    MAP3S sulfate monitoring, 78,
        80–82
    NADP/NTN sulfate
        reductions, 78, 82–87
    NAMS/SLAMS ozone
        monitoring, 55
    nitrogen dioxide respiratory
        illness, 148
    passive smoking, 146–148
    rural ozone, 15
    Salt Lake County
        particulate–mortality, 113
    snow water equivalent, 152
    tetrachloroethylene
        exposure-response,
            121–122, 134
        database, 124
Subset selection, 63–65
    lasso procedure, 64
    leaps procedure, 64
    in DI, 65, 182
    regression, 64
Sulfate concentration
    annual, 83
    daily, 81
    weekly, 81–82
Sum of squares
    error, 10
    penalized, 167
SUM06, 29, 36, 56
SUM08, 29, 36, 56
Surface fitting
    via FUNFITS, 160
Surface-level ozone, 2

Tetrachloroethylene, *see Studies*
Thin-plate spline, *see Spline*
    function

Threshold
> between particulates and mortality, 100

Trend analysis, 10–11
> annual sulfate concentrations, 83–84
> Birmingham mortality, 96
> exceedances, 19
> ozone concentrations, 18
> ozone seasonality, 46
> pollutant monitoring, 74
> semiparametric model, 16

Variable selection
> in mortality analysis, 100
>> backwards elimination, 101, 109

Varimax criterion, *see Principal components analysis*

Variogram
> from FUNFITS vgram, function, 171

Volatile organic compounds, 5

Web companion, 4, 8, 29, 55, 78, 94, 123, 134, 144

Lecture Notes in Statistics

For information about Volumes 1 to 57 please contact Springer-Verlag

Vol. 58: O.E. Barndorff-Nielsen, P. Blaesild, P.S. Eriksen, Decomposition and Invariance of Measures, and Statistical Transformation Models. v, 147 pages, 1989.

Vol. 59: S. Gupta, R. Mukerjee, A Calculus for Factorial Arrangements. vi, 126 pages, 1989.

Vol. 60: L. Gyorfi, W. Härdle, P. Sarda, Ph. Vieu, Nonparametric Curve Estimation from Time Series. viii, 153 pages, 1989.

Vol. 61: J. Breckling, The Analysis of Directional Time Series: Applications to Wind Speed and Direction. viii, 238 pages, 1989.

Vol. 62: J.C. Akkerboom, Testing Problems with Linear or Angular Inequality Constraints. xii, 291 pages, 1990.

Vol. 63: J. Pfanzagl, Estimation in Semiparametric Models: Some Recent Developments. iii, 112 pages, 1990.

Vol. 64: S. Gabler, Minimax Solutions in Sampling from Finite Populations. v, 132 pages, 1990.

Vol. 65: A. Janssen, D.M. Mason, Non-Standard Rank Tests. vi, 252 pages, 1990.

Vol 66: T. Wright, Exact Confidence Bounds when Sampling from Small Finite Universes. xvi, 431 pages, 1991.

Vol. 67: M.A. Tanner, Tools for Statistical Inference: Observed Data and Data Augmentation Methods. vi, 110 pages, 1991.

Vol. 68: M. Taniguchi, Higher Order Asymptotic Theory for Time Series Analysis. viii, 160 pages, 1991.

Vol. 69: N.J.D. Nagelkerke, Maximum Likelihood Estimation of Functional Relationships. V, 110 pages, 1992.

Vol. 70: K. Iida, Studies on the Optimal Search Plan. viii, 130 pages, 1992.

Vol. 71: E.M.R.A. Engel, A Road to Randomness in Physical Systems. ix, 155 pages, 1992.

Vol. 72: J.K. Lindsey, The Analysis of Stochastic Processes using GLIM. vi, 294 pages, 1992.

Vol. 73: B.C. Arnold, E. Castillo, J.-M. Sarabia, Conditionally Specified Distributions. xiii, 151 pages, 1992.

Vol. 74: P. Barone, A. Frigessi, M. Piccioni, Stochastic Models, Statistical Methods, and Algorithms in Image Analysis. vi, 258 pages, 1992.

Vol. 75: P.K. Goel, N.S. Iyengar (Eds.), Bayesian Analysis in Statistics and Econometrics. xi, 410 pages, 1992.

Vol. 76: L. Bondesson, Generalized Gamma Convolutions and Related Classes of Distributions and Densities. viii, 173 pages, 1992.

Vol. 77: E. Mammen, When Does Bootstrap Work? Asymptotic Results and Simulations. vi, 196 pages, 1992.

Vol. 78: L. Fahrmeir, B. Francis, R. Gilchrist, G. Tutz (Eds.), Advances in GLIM and Statistical Modelling: Proceedings of the GLIM92 Conference and the 7th International Workshop on Statistical Modelling, Munich, 13-17 July 1992. ix, 225 pages, 1992.

Vol. 79: N. Schmitz, Optimal Sequentially Planned Decision Procedures. xii, 209 pages, 1992.

Vol. 80: M. Fligner, J. Verducci (Eds.), Probability Models and Statistical Analyses for Ranking Data. xxii, 306 pages, 1992.

Vol. 81: P. Spirtes, C. Glymour, R. Scheines, Causation, Prediction, and Search. xxiii, 526 pages, 1993.

Vol. 82: A. Korostelev and A. Tsybakov, Minimax Theory of Image Reconstruction. xii, 268 pages, 1993.

Vol. 83: C. Gatsonis, J. Hodges, R. Kass, N. Singpurwalla (Editors), Case Studies in Bayesian Statistics. xii, 437 pages, 1993.

Vol. 84: S. Yamada, Pivotal Measures in Statistical Experiments and Sufficiency. vii, 129 pages, 1994.

Vol. 85: P. Doukhan, Mixing: Properties and Examples. xi, 142 pages, 1994.

Vol. 86: W. Vach, Logistic Regression with Missing Values in the Covariates. xi, 139 pages, 1994.

Vol. 87: J. Müller, Lectures on Random Voronoi Tessellations.vii, 134 pages, 1994.

Vol. 88: J. E. Kolassa, Series Approximation Methods in Statistics. Second Edition, ix, 183 pages, 1997.

Vol. 89: P. Cheeseman, R.W. Oldford (Editors), Selecting Models From Data: AI and Statistics IV. xii, 487 pages, 1994.

Vol. 90: A. Csenki, Dependability for Systems with a Partitioned State Space: Markov and Semi-Markov Theory and Computational Implementation. x, 241 pages, 1994.

Vol. 91: J.D. Malley, Statistical Applications of Jordan Algebras. viii, 101 pages, 1994.

Vol. 92: M. Eerola, Probabilistic Causality in Longitudinal Studies. vii, 133 pages, 1994.

Vol. 93: Bernard Van Cutsem (Editor), Classification and Dissimilarity Analysis. xiv, 238 pages, 1994.

Vol. 94: Jane F. Gentleman and G.A. Whitmore (Editors), Case Studies in Data Analysis. viii, 262 pages, 1994.

Vol. 95: Shelemyahu Zacks, Stochastic Visibility in Random Fields. x, 175 pages, 1994.

Vol. 96: Ibrahim Rahimov, Random Sums and Branching Stochastic Processes. viii, 195 pages, 1995.

Vol. 97: R. Szekli, Stochastic Ordering and Dependence in Applied Probability. viii, 194 pages, 1995.

Vol. 98: Philippe Barbe and Patrice Bertail, The Weighted Bootstrap. viii, 230 pages, 1995.

Vol. 99: C.C. Heyde (Editor), Branching Processes: Proceedings of the First World Congress. viii, 185 pages, 1995.

Vol. 100: Wlodzimierz Bryc, The Normal Distribution: Characterizations with Applications. viii, 139 pages, 1995.

Vol. 101: H.H. Andersen, M.Højbjerre, D. Sørensen, P.S.Eriksen, Linear and Graphical Models: for the Multivariate Complex Normal Distribution. x, 184 pages, 1995.

Vol. 102: A.M. Mathai, Serge B. Provost, Takesi Hayakawa, Bilinear Forms and Zonal Polynomials. x, 378 pages, 1995.

Vol. 103: Anestis Antoniadis and Georges Oppenheim (Editors), Wavelets and Statistics. vi, 411 pages, 1995.

Vol. 104: Gilg U.H. Seeber, Brian J. Francis, Reinhold Hatzinger, Gabriele Steckel-Berger (Editors), Statistical Modelling: 10th International Workshop, Innsbruck, July 10-14th 1995. x, 327 pages, 1995.

Vol. 105: Constantine Gatsonis, James S. Hodges, Robert E. Kass, Nozer D. Singpurwalla(Editors), Case Studies in Bayesian Statistics, Volume II. x, 354 pages, 1995.

Vol. 106: Harald Niederreiter, Peter Jau-Shyong Shiue (Editors), Monte Carlo and Quasi-Monte Carlo Methods in Scientific Computing. xiv, 372 pages, 1995.

Vol. 107: Masafumi Akahira, Kei Takeuchi, Non-Regular Statistical Estimation. vii, 183 pages, 1995.

Vol. 108: Wesley L. Schaible (Editor), Indirect Estimators in U.S. Federal Programs. viii, 195 pages, 1995.

Vol. 109: Helmut Rieder (Editor), Robust Statistics, Data Analysis, and Computer Intensive Methods. xiv, 427 pages, 1996.

Vol. 110: D. Bosq, Nonparametric Statistics for Stochastic Processes. xii, 169 pages, 1996.

Vol. 111: Leon Willenborg, Ton de Waal, Statistical Disclosure Control in Practice. xiv, 152 pages, 1996.

Vol. 112: Doug Fischer, Hans-J. Lenz (Editors), Learning from Data. xii, 450 pages, 1996.

Vol. 113: Rainer Schwabe, Optimum Designs for Multi-Factor Models. viii, 124 pages, 1996.

Vol. 114: C.C. Heyde, Yu. V. Prohorov, R. Pyke, and S. T. Rachev (Editors), Athens Conference on Applied Probability and Time Series Analysis Volume I: Applied Probability In Honor of J.M. Gani. viii, 424 pages, 1996.

Vol. 115: P.M. Robinson, M. Rosenblatt (Editors), Athens Conference on Applied Probability and Time Series Analysis Volume II: Time Series Analysis In Memory of E.J. Hannan. viii, 448 pages, 1996.

Vol. 116: Genshiro Kitagawa and Will Gersch, Smoothness Priors Analysis of Time Series. x, 261 pages, 1996.

Vol. 117: Paul Glasserman, Karl Sigman, David D. Yao (Editors), Stochastic Networks. xii, 298, 1996.

Vol. 118: Radford M. Neal, Bayesian Learning for Neural Networks. xv, 183, 1996.

Vol. 119: Masanao Aoki, Arthur M. Havenner, Applications of Computer Aided Time Series Modeling. ix, 329 pages, 1997.

Vol. 120: Maia Berkane, Latent Variable Modeling and Applications to Causality. vi, 288 pages, 1997.

Vol. 121: Constantine Gatsonis, James S. Hodges, Robert E. Kass, Robert McCulloch, Peter Rossi, Nozer D. Singpurwalla (Editors), Case Studies in Bayesian Statistics, Volume III. xvi, 487 pages, 1997.

Vol. 122: Timothy G. Gregoire, David R. Brillinger, Peter J. Diggle, Estelle Russek-Cohen, William G. Warren, Russell D. Wolfinger (Editors), Modeling Longitudinal and Spatially Correlated Data. x, 402 pages, 1997.

Vol. 123: D. Y. Lin and T. R. Fleming (Editors), Proceedings of the First Seattle Symposium in Biostatistics: Survival Analysis. xiii, 308 pages, 1997.

Vol. 124: Christine H. Müller, Robust Planning and Analysis of Experiments. x, 234 pages, 1997.

Vol. 125: Valerii V. Fedorov and Peter Hackl, Model-oriented Design of Experiments. viii, 117 pages, 1997.

Vol. 126: Geert Verbeke and Geert Molenberghs, Linear Mixed Models in Practice: A SAS-Oriented Approach. xiii, 306 pages, 1997.

Vol. 127: Harald Niederreiter, Peter Hellekalek, Gerhard Larcher, and Peter Zinterhof (Editors), Monte Carlo and Quasi-Monte Carlo Methods 1996, xii, 448 pp., 1997.

Vol. 128: L. Accardi and C.C. Heyde (Editors), Probability Towards 2000, x, 356 pp., 1998.

Vol. 129: Wolfgang Härdle, Gerard Kerkyacharian, Dominique Picard, and Alexander Tsybakov, Wavelets, Approximation, and Statistical Applications, xvi, 265 pp., 1998.

Vol. 130: Bo-Cheng Wei, Exponential Family Nonlinear Models, ix, 240 pp., 1998.

Vol. 131: Joel L. Horowitz, Semiparametric Methods in Econometrics, ix, 204 pp., 1998.

Vol. 132: Douglas Nychka, Walter W. Piegorsch, and Lawrence H. Cox (Editors), Case Studies in Environmental Statistics, viii, 200 pp., 1998.